W9-AED-912

The Interpretation of Infrared Spectra

A Programmed Introduction

HEYDEN PROGRAMMED TEXTS

Infrared Spectroscopy

A Programmed Introduction to Infrared Spectroscopy

B.W. Cook and K. Jones

For complementary information on instrumental techniques, the principles and practice of infrared spectroscopy.

Nuclear Magnetic Resonance Spectroscopy

The Interpretation of Proton Magnetic Resonance Spectra

E.J. Haws, R.R. Hill and D.J. Mowthorpe

Six graduated stages lead the beginner to a full appreciation of the assignment of structure to particular features of proton magnetic resonance spectra.

Gas-Liquid Chromatography

A Programmed Introduction to Gas-Liquid Chromatography

2nd Edition **J.B. Pattison**

The book teaches gas-liquid chromatography in a logical progression, and is also designed to be used as a reference work and fault finder in everyday problem solving.

Inorganic Chemistry

Fundamentals of Inorganic Chemistry

Edited by B.J. Aylett and D.E. Billing

A set of 7 programmes which together form a complete teaching package, designed to complement lectures at degree level.

HEYDEN AUDIO-VISUAL PRODUCTS

Interpretation of Infrared Spectra—An Audio-Visual Course

J.H. van der Maas with E.T.G. Lutz **Consultant: H.A. Willis**

A carefully devised programme of 122 colour slides, plus 4 magnetic tape cassettes, course books, correlation charts and answers, leading the student, either studying individually or in a class situation, through a course of 8 lessons, to an understanding of infrared bands characteristic of commonly occurring functional groups, and thereby to the deduction of structural features of compounds.

The Interpretation of Infrared Spectra

A Programmed Introduction

R. R. Hill

The Open University, Milton Keynes

D. A. E. Rendell

The Polytechnic, Wolverhampton

London · New York · Rheine

Heyden & Son Ltd., Spectrum House, Alderton Crescent, London NW4 3XX.
Heyden & Son Inc., 225 Park Avenue, New York, N.Y. 10017, U.S.A.
Heyden & Son GmbH, 4440 Rheine/Westf., Münsterstrasse 22, Germany.

ISBN 0 85501 066 5

Printed in Great Britain by J. W. Arrowsmith Ltd., Bristol BS3 2NT

Contents

Preface

Infrared (i.r.) spectroscopy is now well established as a basic technique for the qualitative and quantitative analysis of organic substances. The infrared spectrophotometer is standard hardware in very many research, analytical and quality control laboratories. In recent years, it has also become a part of the undergraduate teaching laboratory. In all these situations, the major assistance offered by i.r. spectroscopy comes from its empirical use. The documentation on infrared absorption frequencies is now so immense that the interpretation of spectra is frequently simply a matter of matching frequencies with those given for a wide range of atomic groupings in correlation charts. Even this is unnecessary for quantitative analysis, where the strength of a given feature in the spectrum is taken as a measure of the amount of substance present. If this empirical approach solved all problems or gave the maximum possible information, the use of infrared spectroscopy would become a purely mechanical operation.

Nearly every practising organic chemist, however, has met situations where the correlation charts, however elaborate, have been inadequate, or where he has been able to use his knowledge of principles of infrared absorption to obtain further useful information about a substance from its spectrum. Besides this, most teachers have seen students misled by attaching too much significance to unreliable absorption features. Furthermore, when it comes to evaluating the technique for a particular problem, one generally has to draw on a broader knowledge of its potential than that obtainable from correlation charts. It is for these and other reasons that intending users of i.r. spectroscopy are well-advised to become familiar with the principles of the subject as well as the potential of purely empirical methods.

The general aim of this book is to combine effective teaching of basic principles and the use of empirical data with a means of rapid accumulation of experience in interpreting spectra. By using a linear programmed style we are able to supply a continuous feed-back to the student on his understanding of the material and also provide step-by-step guidance in the interpretation of spectra.

The book is divided into five parts, each a self-contained teaching 'package', but assuming a knowledge of what is contained in the previous parts. The first part presents a non-mathematical introduction to i.r. radiation and its units of 'measurement', the interaction of i.r. radiation with matter and the principles of absorption. Molecular vibrations are also introduced in Part 1, followed by the origin of absorption spectra. Part 2 covers experimental aspects in so far as they affect interpretation and choice of technique. It is not a set of practical instructions. The empirical approach is introduced in Part 3 and the idea of characteristic group frequencies established. The main theme is the association of absorption features with molecular fragments. The complementary process of predicting infrared absorption characteristics from structural information is the subject of Part 4 and all the aspects in the programme are integrated in the final part which is concerned with techniques of interpretation. In addition to the programmed text, Parts 1, 3 and 4 also include the following:

(i) a statement of aims, new terms and concepts and assumed prior knowledge;
(ii) a list of objectives;
(iii) an introduction which forms a link with the previous part (except in the case of Part 1);
(iv) a (pre-) self-assessment test;
(v) remarks on pre-test scores;
(vi) summary;
(vii) a (post-) self-assessment test;
(viii) remarks on post-test scores;
(ix) a bibliography;
(x) problems;
(xi) a link paragraph to the next part.

There are five correlation charts on a sheet inserted in the pocket at the back of the book. These can easily be displayed while working through the programme, and kept with the book they are a permanent source of reference when using infrared analysis in the laboratory.

Infrared spectroscopy is rarely used on its own in solving a problem but as part of an iterative process involving other physical and chemical data. We take this into account in selecting problems and where, as in most instances, we assume no prior knowledge of other techniques, the appropriate conclusions one could draw from these are given to the student.

We hope the book will be of value to a wide range of chemists and chemistry students. The overall aim of the book is to enable the reader to become competent in the interpretation of infrared spectra and to be able to recognise situations where i.r. spectroscopy could be used with advantage. It is particularly directed to the student and we have attempted to adapt it for classroom use. It is not necessary that the whole programme should be used at once. In particular, Parts 4 and 5 can be taken after a break from Part 3. The reader already familiar with infrared spectroscopy could use Parts 1, 2 and 3 as revision material, and Parts 4 and 5 to gain a better understanding of more detailed aspects of functional group absorption characteristics.

In the classroom, the programme can be used as a component in a course in i.r. spectroscopy. The main facts and concepts can be left to the programme, freeing the teacher to concentrate on applications in tutorials or cope with individual students' queries. The programme allows for different rates of progress; slow students can be assisted by the teacher and, if necessary, encouraged to catch up with the average worker in their own time. Students who make rapid progress can be referred to the questions and the bibliography at the end of each part. The programme can also be used for revision purposes or as an aid to the student who has missed several weeks' work.

We also hope this book will be of considerable help to those who cannot conveniently benefit from classroom/laboratory instruction and have to study or catch up in isolation. Such would be the case of an individual who finds that he has to use i.r. data or is unsure whether it can be used with advantage for his particular problem. The book could be used as an up-dating or revision exercise, or perhaps as an introduction to the subject for the first time. It is largely self-teaching for both kinds of use but readers need to have some knowledge of basic chemistry before starting Part 1.

Programmed Learning

Although interpretive spectroscopy lends itself to programmed learning this is only part of the reason why we chose to use it in our book. It is now becoming widely recognized that the teaching effectiveness of written material can be significantly improved if it is made student-active rather than passive, and if the student can obtain feed-back while he is working, and thereby continuously assess his progress.

Programmed learning encourages this to happen. The subject matter is divided up into units known as frames. At the end of each frame the student finds out immediately whether he has understood the point being made. If he has not, he can rectify matters there and then by re-reading the material more carefully and taking note of any remedial comments. Nevertheless, though it is more self-teaching than most conventional books, a linear programme can never be regarded as a satisfactory medium on its own. This is because it normally allows little room for the student to develop his own thoughts on the subject matter. It is a means of rapidly acquiring a basic knowledge and understanding of a topic which can then be applied in the less intellectually restricted atmosphere of tutorial work and discussion with one's colleagues or fellow students.

This programme is different in many respects from earlier published programmes. We have in the main avoided very short frames, that is to say, frames of low intellectual content. The type of response is highly varied and we have avoided completely the 'missing word' type. We have found that one consequence of using linear programmed style (which guides the students' thoughts step by step through conceptually difficult areas of the topic) is that it has been necessary to introduce generalities at the expense of rigour in one or two places. We felt this was justified where a lengthy digression for the sake of precision would be an undesirable interruption in the main argument.

Finally, the main characteristic of programmed material is that it has been tested and proved to be effective. This book has undergone an objective validation programme and has been tested and revised three times. The evaluation was carried out with the cooperation of thirty college and industry-based students involving widely different academic backgrounds. Students used the criterion tests associated with Parts 1, 3 and 4 to assess their ability to achieve the objectives of each part before and after working through the relevant material. The final version of the programme enabled the average student with little or no knowledge of the subject to increase his test performance as follows:

	Pre-test score (%)	*Post-test score* (%)
Part 1	30	90
Part 3	15	85
Part 4	40	90

The validation was only possible through the voluntary help of many students and their supervisors in the ten institutions which participated. We gratefully acknowledge the considerable time and effort they spent on the manuscript and the constructive comments they offered for its improvement. Drs. R. W. Alder and T. Threlfall made particularly detailed comments and valuable suggestions, as did many of our colleagues at Wolverhampton Polytechnic.

September, 1974

R. R. Hill
D. A. E. Rendell

How to Use the Programme

You will probably find the best approach to each part (except Part 2 which is not programmed) is as follows. Take note of assumed prior knowledge, read the introduction and attempt the self-assessment test. The test is important because it gives you a more objective view of your present knowledge of the material and, by comparing your first score with the score you obtain after working through the subsequent programmed material, you will have direct evidence of what has been achieved. The programme itself is written in a fairly common format and we suggest you read it in the following way:

(i) Cover the page with a sheet of paper and pull it down to the first line drawn across the page.

(ii) Read the material exposed and attempt the question (or in some cases, choose the most appropriate word from those printed in italic type), writing your answer on a separate sheet of paper.

(iii) Pull the covering sheet down to the next full line, revealing the next frame and the answer to the first.

(iv) Check your answer to frame 1 with that given.

  (a) If it is correct read the material of frame 2 and continue as from (ii) above.

  (b) If it is incorrect take note of any remedial comment given with the answer or re-read frame 1 and try to understand the reason for your incorrect response. Then continue as from (ii) above.

When you have worked through the programmed material in a part, read the summary and attempt the self-assessment test, marking your answers as before.

The continuous activity in programmed instruction is necessarily more demanding than reading a normal text and you should not study for more than about 40 minutes without some kind of a break. Each part of the programme has been divided into sections for this purpose though you may find you can cover more than one section at a sitting. On the other hand the overall aim of the programme will not be achieved if you spread the work on either of the two halves (Parts 1, 2 and 3, or Parts 4 and 5) over too long a period. You should aim to complete each of these halves within a period of less than three months. The average times spent on each section by students who validated the programme are given on p. xii. We hope these figures will be helpful in planning your work.

However the programme is used, the retention of the subject matter will depend very much on how frequently you use i.r. spectroscopy afterwards. As with proficiency in languages, continuous practice is necessary not only for improvement but for maintaining a level of competence. In this regard it is perhaps worthwhile noting that much intellectual satisfaction can be had from practising one's skill on problems in spectroscopy. Moreover the solving of such puzzles as are presented in books given in the bibliography for Part 5 can be not only a useful activity but an enjoyable one as well.

Guide to Programme Timing

Here as a guide are the times in minutes spent on average by the students who helped to validate the book.

PART 1	Electromagnetic Radiation and its Parameters	30
	The Interaction of Infrared Radiation with Matter	30
	Molecular Vibrations	25
	Infrared Absorption Spectra	40
	Overtones and Combination Bands	20
PART 2		35
PART 3	Empirical Use of Frequencies	25
	Single Bond Stretching Vibrations	15
	Multiple Bond Stretching Vibrations	10
	Skeletal Vibrations	30
	Characteristic Frequencies below 1500 cm^{-1}	30
	Absorption Characteristics of the Carbonyl Group	35
	The Elucidation of Structure from Infrared Spectra	35
PART 4	Saturated Hydrocarbons	25
	Unsaturated Hydrocarbons	25
	Alcohols, Phenols, Amines and Ethers	20
	Ketones and Aldehydes	20
	Carboxylic Acids	20
	Carboxylic Esters and Lactones	20
	Acid Halides, Anhydrides and Amides	25
	Other Nitrogen-containing Groups	20
	Sulphur- and Halogen-containing Compounds	20
PART 5	Each problem took approximately 20 minutes.	

Topic Network

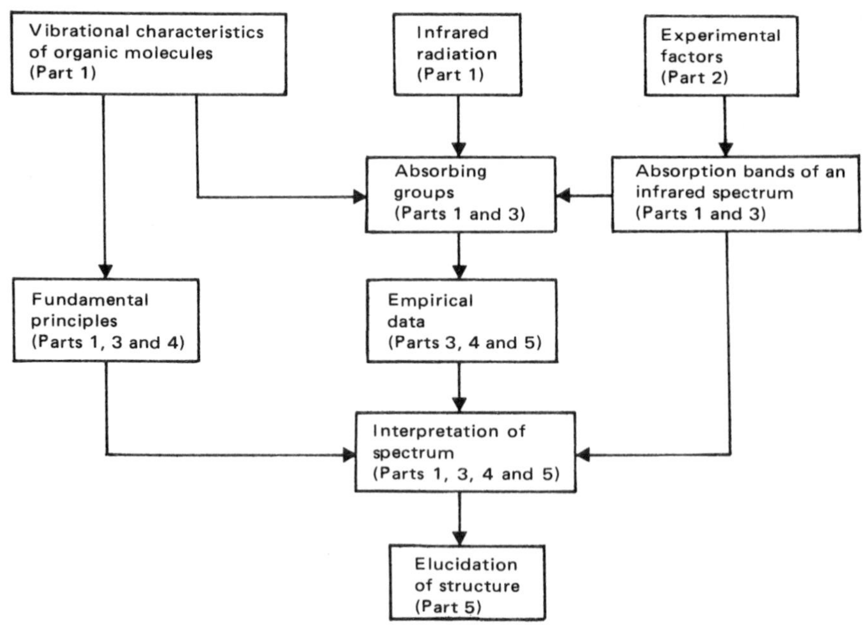

PART 1
BASIC PRINCIPLES

Aim

The intention of the first part is to cover the fundamental principles behind infrared spectroscopy and take the student up to the point where he understands the origin of the features in an infrared spectrum. This is done by describing

(i) electromagnetic radiation and the parameters associated with it;
(ii) the interaction of radiation with matter and the correlation of the absorption of radiation in various regions of the spectrum with different types of molecular energy levels;
(iii) physical pictures of molecular vibrations;
(iv) factors which limit the number of absorption bands in a spectrum; and
(v) features of infrared spectra which are not due to fundamental vibrations.

New Terms and Concepts

Electromagnetic radiation, wavelength, frequency, micrometre (micron), wavenumber, $E = h\nu$, energy level, ground state, excited state, absorption, transition, force constant, mode of vibration, stretching and bending vibration, in-phase, out-of-phase, symmetric, antisymmetric, deformation, rock, wag, twist, absorption band, spectrum, selection rules, fundamental, overtones, combination bands, Fermi resonance.

Prior Knowledge

Elementary aspects of molecular structure, secondary school physics and chemistry, and elementary mathematical operations.

Objectives for Part 1

When you have completed Part 1 you should be able to:

1. Give the correct relationship between any two of the following: wavenumber, frequency, wavelength and energy.

2. Select from a given list of wavelengths, frequencies or wavenumbers those which could induce a transition between the vibrational energy levels of a molecule.

3. Identify aspects of molecular structure responsible for the absorption of infrared radiation.

4. Recognize whether or not a simple vibration involves a change in dipole moment.

5. Recognize vibrations which do and those which do not give rise to absorption of infrared radiation within the normal range of wavenumbers used in infrared spectroscopy.

6. Apply a knowledge of the influence of

   (i) bond order
   (ii) relative atomic masses
   (iii) bond polarity

   on the wavenumbers of absorptions due to stretching vibrations to rank closely related bonds in order of increasing strength or stretching frequency.

7. Classify a particular absorption as a fundamental, an overtone, a combination band or a consequence of Fermi resonance by studying a given selection of the wavenumbers of other strong bands in the spectrum.

PART 1: BASIC PRINCIPLES

Introduction

Infrared spectroscopy is one of the many valuable physical methods currently available to the organic chemist which are based on the interaction of electromagnetic radiation with matter. In the first part of this programme, you will obtain a superficial understanding of the nature of electromagnetic radiation and, in particular, how infrared radiation interacts with molecules. You will become familiar with the parameters and units used to characterize radiation and the way in which they are related to the 'mechanical' properties of molecules. The interpretation of infrared spectra depends on a thorough understanding of the types of vibration which may occur within a molecule. The latter section of this part will describe the vibrational motions involved in some detail, and you will learn to use the specialized terminology associated with this topic. A close study of some very simple molecules will help you later in the programme to put the information to be gained from a study of infrared spectra on a rational basis.

Multiple Choice Test

Before proceeding with the programme itself, attempt the multiple-choice test which follows. In each case only one of the alternatives is correct. Determine which this is and write down its letter (a, b, or c) in a vertical column on a separate sheet of paper. Marking instructions will follow the test.

1. Which of the following relationships is correct?

   (a) $\bar{\nu} = \frac{c}{\lambda}$ (b) $\lambda = \frac{1}{\nu}$ (c) $E = \frac{hc}{\lambda}$

   where h = Planck's constant, c = velocity of light, λ = wavelength, ν = frequency, $\bar{\nu}$ = wavenumber, E = energy of transition.

2. Absorption of infrared radiation by a molecule involves

   (a) electronic, vibrational and rotational energy levels;
   (b) vibrational and rotational energy levels;
   (c) rotational energy levels only.

3. A transition between vibrational levels of a molecule could be responsible for an absorption band occuring at

   (a) 2380 cm^{-1}; (b) 420 μm; (c) 3.8×10^{10} Hz.

4. Which of the following statements is correct?

   (a) Wavenumber is a unit of frequency.
   (b) The energy of radiation is directly proportional to its wavenumber.
   (c) Planck's constant occurs in the formula which relates vibration frequency to the masses of the atoms vibrating.

5. Select the correct statement.
During the vibration illustrated,

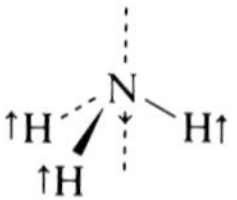

(a) there is no change in dipole moment,
(b) the dipole moment changes its direction but not its magnitude,
(c) the dipole moment changes its magnitude in the direction of the dashed vertical line.

6. Which one of the three vibrations depicted below would not absorb within the normal range of wavenumbers used in infrared spectroscopy?

(a) (b) (c)

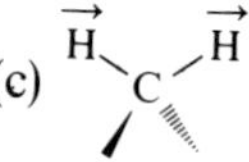

7. The stretching frequency of the carbonyl group is near 1700 cm^{-1}. The C—O stretching frequency of ethers is expected

(a) near 1200 cm^{-1}; (b) near 2000 cm^{-1};
(c) to have approximately the same value.

8. Generally speaking, C=N stretching bands are expected to be

(a) of about the same intensity (b) considerably more intense than
(c) considerably weaker than

C=C stretching absorptions.

9. The vibration of highest frequency in the chloroform molecule will be

(a) the C—H stretch; (b) the H—C—Cl bend;
(c) the C—Cl stretch.

10. An infrared spectrum shows a weak band at 1900 cm^{-1} which does not originate from a fundamental vibration. There are very strong bands in the same spectrum at 620, 950, 1010 and 1375 cm^{-1}. Therefore, the weak band is almost certainly

(a) an overtone;
(b) a combination band;
(c) a consequence of Fermi resonance.

Comments on marks

Mark your sheet as follows: Compare your answer column with the vertical sequence of letters given for this test on page 190. For each question, give yourself +1 mark if you have chosen the correct answer, $-\frac{1}{2}$ mark if you have selected either of the other two and 0 if you did not answer the question. Total your marks for the complete test and compare the result with the following remarks. Do not seek out the correct answer to an incorrectly answered question, unless your mark falls into the third category (below).

Less than 3: Proceed with the programme.

3—6½ You know something about infrared spectroscopy already. However, your knowledge and understanding is not sufficient to ensure a full grasp of what is to come in subsequent parts. Work through this part—it shouldn't take you long.

7 and over Your knowledge of basic principles is adequate to continue directly with Part 2. However, you should check up on those points which led you to a wrong answer in the test.

Electromagnetic Radiation and Its Parameters

F1 Many physical methods employed in organic chemistry involve the interaction of electromagnetic radiation with matter. It is the purpose of this section to explain what kind of interaction is involved in the case of infrared spectroscopy. In order to do this we must first consider the nature of electromagnetic radiation itself. The most familiar example of electromagnetic radiation is light. The nature of light can be adequately described for our purpose as a wave motion, characterized by its wavelength, λ. This concept is most easily understood by considering the analogy of a wave system in water. Such a system is the series of ripples which travel outwards from a central disturbance caused for example by dropping a stone into a pond. Each ripple consists of a crest and a trough and travels at a velocity of a few centimetres per second. The wavelength λ of the wave motion is the distance between two consecutive crests, and in the water example amounts to a few centimetres (Fig. 1.1).

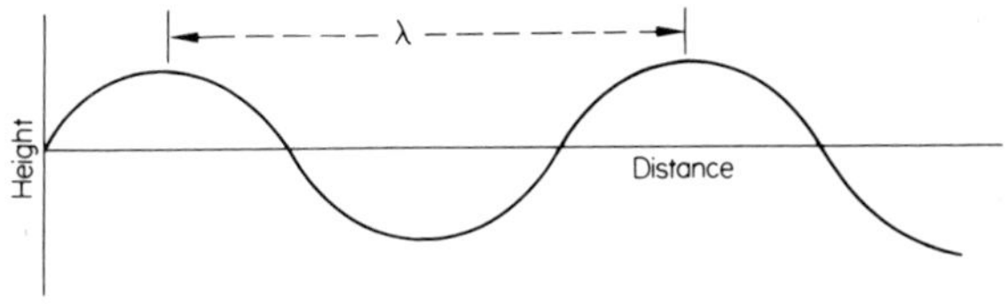

Fig. 1.1

What is the distance between a crest and a neighbouring trough in a wave motion of wavelength 10 cm?

Express this also in terms of λ.

F2 As an alternative to measuring the wavelength we could also characterize the wave motion by counting the number of waves that pass a stationary reference point in a given time. This quantity is the frequency and it is given the symbol ν. The variation of the height of the surface of the water at the reference point with time may be shown diagrammatically (Fig. 1.2).

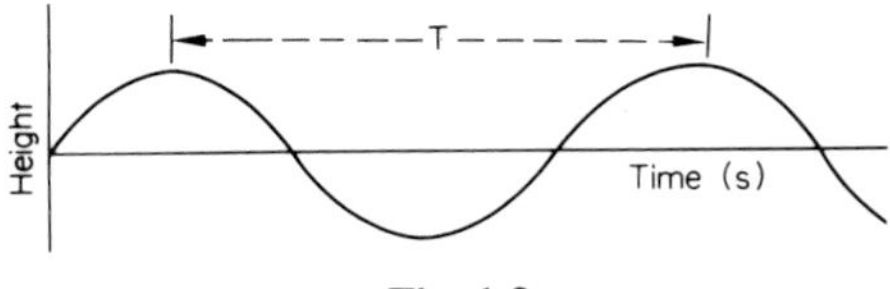

Fig. 1.2

The motion undergoes one complete cycle in the time T and hence the frequency ν is the number of *cycles per second/seconds in one cycle.* (Select the correct term of those italicized and separated by an oblique stroke. This convention for multiple choice alternatives is used frequently throughout the programme.)

A1 5 cm, $\lambda/2$ (see Fig. 1.1).

F3 The unit 'cycle per second' is now known as the 'Hertz'.

$$1 \text{ cycle per second} = 1 \text{ Hertz (Hz)}$$

The wavelength and the frequency are connected by the formula

$$u = \lambda v$$

where u is the velocity of propagation of the wave. Thus, for a ripple system in water travelling at 20 cm s^{-1} and having a wavelength of 5 cm the frequency is *100*/*4*/*0.25* Hz.

A2 cycles per second. Frequency = no. of waves passing a fixed point in one second.

F4 In the case of visible light the height of the wave in Fig. 1.1 and Fig. 1.2 becomes the magnitude of an electromagnetic field. Figure 1.3 and Fig. 1.4 are analogous to Fig. 1.1 and Fig. 1.2 but refer to an electromagnetic wave. Suggest terms for the ordinates y and p and the marked quantities x and q.

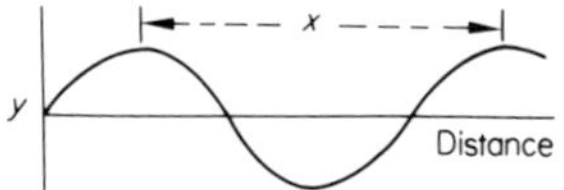

Fig. 1.3

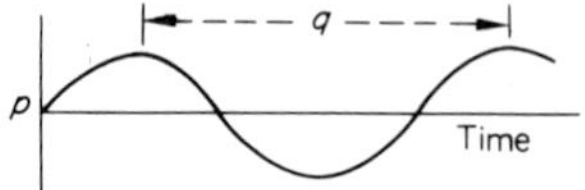

Fig. 1.4

A3 4 Hz $\quad v = u/\lambda = 20/5$.

THE ELECTROMAGNETIC WAVE IS A FORM OF ENERGY—we shall refer to this later.

F5 Open out the folded sheet in the pocket at the back of the book. Refer to Chart 1 for the following twenty frames. Here we see that visible light is only a very small part of the electromagnetic spectrum which extends from very short wavelength (X-rays) to very long wavelength (radio waves).

Infrared radiation has a *longer*/*shorter* wavelength and covers a *larger*/*smaller* range than visible light.

A4 Figure 1.3: ordinate y = electromagnetic field; x = wavelength. Figure 1.4: ordinate p = electromagnetic field; q = one cycle.

F6 Which of the following statements is/are true?

(a) Electromagnetic radiation can be regarded as a form of energy covering a wide range of wavelengths and frequencies.
(b) Radio waves are of higher frequency than X-rays.
(c) Radio waves are of longer wavelength than X-rays.

A5 longer, larger.

F7 Electromagnetic radiation of all wavelengths travels at the same velocity, c, in a vacuum. This is a well-known physical constant most frequently referred to as the velocity of light. Write down an expression which relates the wavelength of light to its frequency and velocity.

A6 (a) and (c) are true.

F8 In the case of visible light, the wavelength is very short, ranging from about 4×10^{-5} cm for violet light to 7.5×10^{-5} cm for red light. In order to avoid using such small numbers, it is convenient to use the smaller subdivisions of the metric system such as the 'micrometre', μm (formerly known as the micron).

$$1\ \mu\text{m} = 10^{-6}\ \text{m} = 10^{-4}\ \text{cm}$$

In terms of this unit, the wavelength of visible light ranges from _____ to _____ μm.

A7 $c = \nu\lambda$.

F9 The 'middle' infrared region, with which we shall be concerned, covers the range 2 to 25 μm. This is equivalent to _____ to _____ cm.

A8 0.400 to 0.750 μm.

F10 We can readily convert wavelengths to frequencies by the expression $c = \nu\lambda$, the numerical value of c being 3.0×10^{10} cm s^{-1}. Calculate the frequencies of the limits of the middle infrared region of the electromagnetic spectrum.

A9 2×10^{-4} to 25×10^{-4} cm.

F11 An alternative way of characterizing electromagnetic radiation is to use the reciprocal of wavelength in centimetres. This quantity, which is the number of waves in 1 cm length, is known as the wavenumber, and given the symbol $\bar{\nu}$. Express $\bar{\nu}$ in terms of λ.

A10 $\nu = c/\lambda$. Therefore ν for the low wavelength limit is given by

$$\frac{3 \times 10^{10}}{2 \times 10^{-4}} = 1.5 \times 10^{14}\ \text{Hz}$$

ν for the high wavelength limit is

$$\frac{3 \times 10^{10}}{25 \times 10^{-4}} = 1.2 \times 10^{13}\ \text{Hz}$$

F12 The unit for the wavenumber is cm^{-1}. Thus, infrared radiation of wavelength 2 μm(2×10^{-4} cm) has a wavenumber of _____ cm^{-1}.

A11 $\bar{\nu} = \dfrac{1}{\lambda}$.

F13 The long wavelength limit of the middle infrared region has a wavenumber of _____ cm^{-1}. (Use previous frames, not Chart 1).

A12 5000 cm^{-1}.

F14 There are three quantities which we can use in the description of electromagnetic radiation. They have the symbols ν, $\bar{\nu}$ and λ.

What are they?

A13 400 cm^{-1}.

F15 Wavelength and wavenumber are used in the description of infrared phenomena. Furthermore modern preference is for the wavenumber.

The limits of the middle infrared, in terms of wavenumber are:

1.5 × *10*^{14}*—2* × *10*^{13}*/5000—400/2—25* cm^{-1},

and in wavelength are:

1.5 × *10*^{14}*—2* × *10*^{13}*/5000—400/2—25* μm.

A14 frequency, wavenumber and wavelength.

F16 Care must be taken not to confuse the symbol ν, the frequency in Hertz, with the symbol $\bar{\nu}$, the wavenumber in cm^{-1}. The relationship between them is obtained from the expressions $\nu\lambda = c$, $\bar{\nu} = 1/\lambda$.

Hence $\bar{\nu} = ?$

A15 5000—400 cm^{-1}, 2—25 μm.

F17 Although, as we have seen in **F15**, the wavenumber is the preferred parameter used in characterizing infrared radiation, it is unfortunately common practice to use the term 'frequency' in its place. This is clearly incorrect but one will often encounter phrases such as 'a *frequency* of 1000 cm^{-1}' in writing and 'a *frequency* of 1000 wavenumbers' in speech, instead of the correct 'a wavenumber of 1000 cm^{-1}'. This formally incorrect usage is now so firmly established that it would be pedantic to insist on the correct phraseology. We shall, in fact, adopt the common usage later on in the programme. Re-write the following sentences in their strictly correct form.

'Cyclohexanone absorbs at a frequency of 1710 cm^{-1}. The corresponding absorption of cyclohexenone is to be found at a lower frequency (1690 cm^{-1})'.

A16 $\bar{\nu} = \nu/c$.

A17 'Cyclohexanone absorbs at a wavenumber of 1710 cm^{-1}. The corresponding absorption of cyclohexenone is to be found at lower wavenumber (1690 cm^{-1})'. Note also that the symbol ν without the bar is sometimes used to stand for the wavenumber in cm^{-1}. The meaning of the symbol, however, is usually clear from the context.

The Interaction of Infrared Radiation with Matter

F18 We now turn to consider the interaction of infrared radiation with matter. It was stated in **F4** that electromagnetic radiation, of which infrared is a small part, is a form of energy. In fact radiation of a particular wavelength or frequency has a unique energy associated with it given by the equation $E = h\nu$ where h is a constant (Planck's constant) and ν is the frequency of the radiation. Therefore, not only can we express a particular form of electromagnetic radiation in terms of wavelength, wavenumber or frequency but, because of this expression we can formulate it in energy terms using any of the energy units in common use e.g. electron volts, Joule $mole^{-1}$ etc. (see Chart 1). Express the energy of electromagnetic radiation, E, in terms of a constant and the wavelength λ.

F19 We have expressed electromagnetic radiation in energy terms, now let us consider the energy structure of matter. According to quantum theory, the energy of a molecule can assume only certain definite values. These values are known as energy levels and are usually represented as horizontal lines on a vertical energy scale to give an energy level diagram. The lowest energy level is known as the ground state, the other levels being known as excited states. A molecule in an excited state therefore has *less/more* energy than a ground state molecule.

A18 $E = h\nu$ and $\nu = c/\lambda$; therefore $E = hc/\lambda$. h is Planck's constant and c, the velocity of light, is constant, hence $E = \text{constant}/\lambda$.

F20 Which of the following statements is/are false?

(a) The energy of electromagnetic radiation is inversely proportional to its frequency.
(b) An energy level diagram can be used to illustrate the process of absorption of energy.
(c) In principle the energy of a molecule can take any value.

A19 more.

F21 When a molecule absorbs energy it undergoes a **transition** from the ground state to an excited state (represented by the arrow in Fig. 1.5). Absorption can

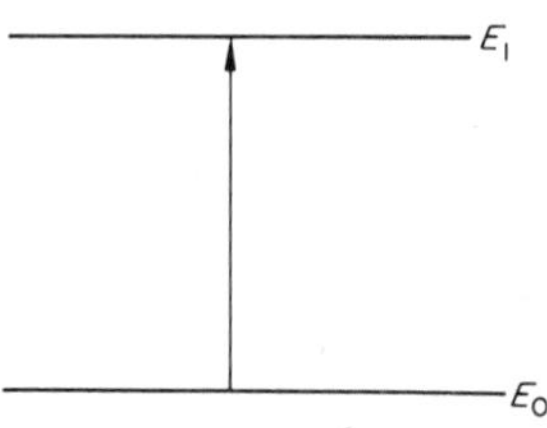

Fig. 1.5

therefore be described as a process in which the interaction of electromagnetic radiation with a molecule causes the molecule to undergo a transition from its original energy level to a *higher/lower* energy level.

A20 (a) and (c) are false. (a) $E = h\nu$; (c) see **F19**.

F22 What does the arrow in Fig. 1.5 represent? (one word answer).

A21 higher.

F23 If a molecule undergoes a transition from the ground state to the first excited state, its energy is increased by $E_1 - E_0$, where E_0 is the energy of the *ground/excited* state (see Fig. 1.5).

A22 transition or absorption.

F24 We have already seen that the energy of electromagnetic radiation of frequency ν is $h\nu$ where h is Planck's constant. Absorption of this radiation will therefore cause a molecule to undergo a transition from the ground state to the first excited state provided that $E_1 - E_0 = ?$

A23 ground.

F25 The energy levels of a molecule fall naturally into three categories:

(i) electronic
(ii) vibrational
(iii) rotational.

A molecule can exist in one of a number of electronic states according to the distribution of its electrons among the available orbitals. The distribution which gives the lowest energy is the electronic ground state of the molecule and all the others are electronic excited states. Each of these electronic states is subdivided into levels of vibrational energy and each vibrational level is subdivided into levels of rotational energy. Thus we can represent the energy levels of a molecule diagrammatically as shown in Fig. 1.6.

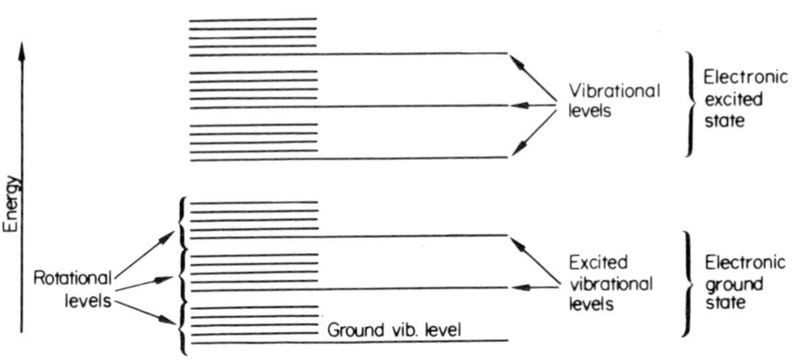

Fig. 1.6

The energy gaps between rotational levels in a molecule are *smaller/larger* than those between vibrational levels which are *smaller/larger* than those between electronic levels.

A24 $h\nu$.

F26 We are concerned in this programme with the absorption of infrared radiation in the range 400—4000 cm^{-1} only. It so happens that the energy of this radiation is of the same order of magnitude as the separation between the VIBRATIONAL levels of a molecule. Consequently the absorption of infrared radiation will increase the vibrational energy of a molecule. Is this last sentence true or false?

A25 smaller, smaller.

F27 Select the correct terms in each of the following sentences:

(a) Infrared radiation is characterized by its *energy level/wavenumber/vibrational energy*.
(b) The correct symbol for wavenumber is $\lambda/v/\bar{v}/E_0$.
(c) The *absorption/transition* of infrared radiation in the range 400—4000 cm^{-1} by a molecule results in an *absorption/transition* between its *electronic/vibrational/rotational* energy levels.

A26 true.

F28 In order to understand more fully the absorption of infrared radiation by molecules, and to interpret the results obtained, it is helpful to consider in detail the vibration of a simple diatomic molecule such as HCl or CO. We can think of this in terms of a mechanical model consisting of two balls connected by a spring. If the spring is extended and suddenly released the 'molecule' will start to vibrate with a frequency, say ω. Such a motion constitutes a stretching vibration, in the course of which at any given instant both balls are moving in the *same direction/opposite directions* and *perpendicular to/along* the 'bond axis'.

A27 (a) wavenumber (**F11**).
(b) $\bar{v}$ (**F14**).
(c) absorption, transition, vibrational (**F21—F26**).

F29 The frequency of vibration of our model will depend on the masses of the balls and the strength of the spring connecting them. Classical mechanics provides an expression for the frequency in terms of these two quantities:

$$\omega = \frac{1}{2\pi}\sqrt{\frac{k}{m_1 m_2/(m_1 + m_2)}}\ \mathrm{Hz}$$

where k can be regarded as a measure of the strength of the spring, known as the **force constant,** and m_1 and m_2 are the masses of the balls. Let us now apply this purely mechanical concept to the molecular case. If the force constant k had the same value in HCl (H, mass 1; Cl, mass 35.5) as in HBr (Br mass 80), which molecule would vibrate with the higher frequency, ω?

A28 opposite directions, along

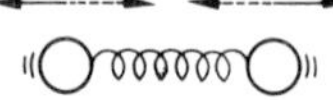

F30 In a comparison between two hypothetical diatomic molecules of the type X—X with the same force constant, one could make the general point that the heavier the atoms, X, the *higher/lower* the molecular vibrational frequency.

A29 $\omega_{HCl} > \omega_{HBr}$. If HCl and HBr had the same force constant, the difference in frequency for the two molecules would depend only on the magnitude of $m_H m_X/(m_H + m_X)$ and examination of the equation in **F29** reveals that the larger this value, the smaller the value of ω. Hence ω_{HCl} will be greater than ω_{HBr}.

F31 Now let us turn to the other variable, and see what happens when we vary the 'strength of the spring', i.e. the force constant. The force constant for a double bond is approximately twice, and a triple bond, three times the value for a single bond between the same atoms. Calculate the approximate vibrational frequencies of the C=N and C≡N groups given that the frequency of the C—N group is 3×10^{13} Hz.

A30 lower. Apply the same reasoning as given in **A29**.

F32 Which of the following statements is/are correct?

The force constant of a bond is

(a) directly proportional to the frequency of vibration.
(b) a measure of the strength of the bond.
(c) larger for a bond between light atoms than one between heavy atoms.

A31 4.24×10^{13} Hz, 5.20×10^{13} Hz, $\omega_2/\omega_1 = \sqrt{(k_2/k_1)}$. $m_1 m_2/(m_1 + m_2)$ is the same for all three groups and cancels out.

F33 It can be shown that the frequency of infrared radiation absorbed by a diatomic molecule is equal to the latter's natural vibration frequency. The situation is very similar to the phenomenon of resonance in acoustics in which a violin string tuned to a frequency of say 500 Hz starts to vibrate when exposed to sound waves of the same frequency. In both cases, energy is transferred from the wave motion to the system exposed to it. The vibrational frequency of HCl is 8658×10^{10} Hz. What frequency of infrared radiation will interact with the HCl molecule?

A32 (b) The force constant is proportional to the **square** of frequency. It is **independent** of the mass of atoms involved.

F34 Infrared energy is transferred to a molecule by a process known as *absorption/transition/vibration.*

A33 8658×10^{10} Hz.

F35 HBr would absorb radiation of *lower/higher* frequency than HCl if the force constants of the two molecules were identical.

A34 absorption.

F36 We have seen that infrared radiation is normally characterized by its wavenumber rather than its frequency. Calculate the wavenumber of radiation which is absorbed by HCl.

A35 lower. Other things being equal, the heavier the atoms the lower the frequency.

A36 2886 cm^{-1}. The vibration frequency of HCl is 8658×10^{10} Hz. Applying

$$\bar{\nu} = \frac{\nu}{c} \text{ (A16)} \qquad \bar{\nu}_{HCl} = \frac{8658 \times 10^{10}}{3 \times 10^{10}} = 2886 \text{ cm}^{-1}$$

Molecular Vibrations

F37 The vibration of a diatomic molecule is easy to envisage and the simple picture gives a complete and rigorous description of the motion. With polyatomic molecules the motion is much more complicated, but can be thought of as a combination of a number of relatively simple vibrations. Each of these, usually termed a mode of vibration, may be considered independently of the others and has a definite vibrational frequency. How many modes of vibration does the HCl molecule have?

F38 Although the total vibrational energy of the molecule is increased by the absorption of infrared radiation of a particular frequency, in effect the energy absorbed is taken up entirely by one mode of vibration which is often localized in one, two, or three bonds. In other words the consequence of absorption in such cases is that the vibrational motion of a *small/large* part of the molecule is increased while the remaining *small/large* part is left relatively unchanged.

A37 One. It can only vibrate by the stretching of the H—Cl bond.

F39 Which of the following statements are true?

(a) The vibration of a polyatomic molecule is more complicated than that of a diatomic molecule.

(b) A molecule having three modes of vibration may absorb i.r. radiation at three different frequencies.

(c) A polyatomic molecule made up of heavy atoms will have lower vibrational frequencies than one composed of light atoms. (Assume similar force constants.)
(d) Energy acquired by the absorption of i.r. radiation is always shared equally amongst the modes of vibration of a polyatomic molecule.

A38 small, large.

F40 Vibrations can either involve changes in bond length (stretching) or in bond angle (bending).

Label the following vibrations as bending or stretching:—

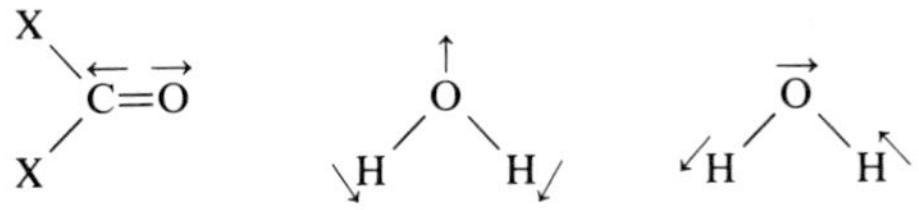

$X_2\overleftarrow{C}{=}\overrightarrow{O}$ H_2O (O ↑, H ↘, H ↙) $H_2\overrightarrow{O}$ (H ↙, H ↖)

A39 (a), (b) and (c) are true.

F41 Draw diagrams to represent the oxygen–hydrogen stretching vibration of an alcohol ROH, the carbon–nitrogen stretching vibration of a cyanide RCN and the bending vibration of the carbon skeleton in propane using the notation given in **F40**.

A40 stretching, bending, stretching. Notice the vibrations of the water molecule are characterized by the arrows on the hydrogen atoms. The arrow on the oxygen atom indicates a very small displacement which is necessary to keep the molecule as a whole in one place. We shall simplify future illustrations by omitting these small displacements.

F42 The stretching of the C—H bonds in the methylene group can be coupled in-phase or out-of-phase. This gives rise to two C—H stretching vibrations in which the length of both C—H bonds alters and which are known as 'symmetrical' and 'antisymmetrical' stretching vibrations.

In-phase symmetric: C with H ↙ and H ↘   Out-of-phase antisymmetric: C with H ↙ and H ↖

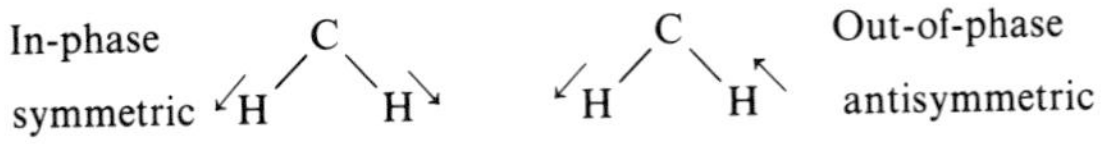

Draw the symmetrical and antisymmetrical stretching vibrations of the amino group, $-NH_2$.

A41 $R\overleftarrow{O}{-}\overrightarrow{H}$ $R\overleftarrow{C}{\equiv}\overrightarrow{N}$ H_3C (↗) — CH_2 (↓) — CH_3 (↖)

F43 On the basis of **F42**, classify the stretching vibration of water shown below.

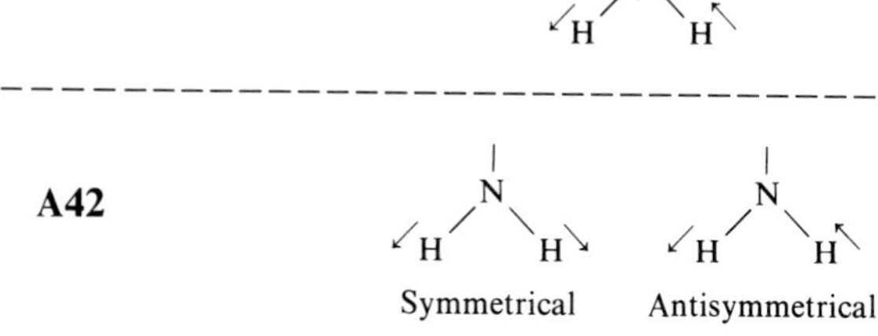

F44 The form of a stretching vibration is fairly straightforward. The description of a mode as, for example, a C—H stretching vibration, qualified by such terms as 'symmetric' and/or 'in-phase' is usually unambiguous. Bending vibrations on the other hand are much less well-defined and they are usually subdivided into 'deformations', 'wags' or 'twists', or similar terms, which approximately describe the motion. Consider the methylene group $—CH_2—$. There are two **in-plane** bending vibrations,

Deformation   Rock

and two **out-of-plane** bending vibrations

Wag   Twist

Thus for the methylene group there are *2/3/4* bending and *1/2/3* stretching vibrations. See if you can draw them all without looking at the programme.

A43 antisymmetric or out-of-phase.

F45 A methylene group in a straight chain molecule such as a normal paraffin has a wagging vibration in which the hydrogen atoms move essentially *perpendicular/parallel* to the direction of the chain.

A44 4, 2.

F46 One of the four bending vibrations of the methylene group occurs **within** the group, while the others involve the movement of the group as a whole. Name the former.

A45 essentially parallel. The movement perpendicular to the chain is a rocking vibration.

F47 Which of the names used to describe the bending vibrations of methylene would be most appropriate for the following bending vibration of the methyl group?

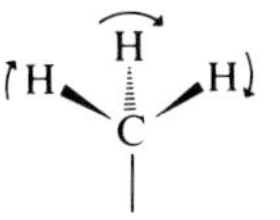

A46 deformation. This is the only vibration which alters the H—C—H angle significantly.

F48 The bending vibration of a single hydrogen atom attached to an unsaturated carbon atom is much simpler than the cases just discussed because it is attached to a much more rigid framework. Consider for example, 2-methyl-2-butene. The vinyl hydrogen atom can undergo a bending vibration, in the plane of the molecule or out of the plane of the molecule. Which is which in the diagrams below?

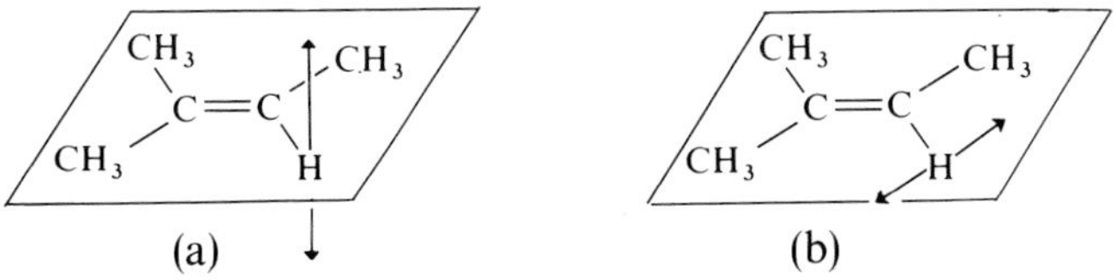

(a) (b)

A47 twist.

F49 Draw the corresponding vibrations for the C—H bond in pentafluorobenzene.

A48 (a) out-of-plane; (b) in-plane.

F50 In the following diagrams, + indicates movement up from the plane of the paper, and – indicates movement down from the plane of the paper. Which of the captions are incorrect?

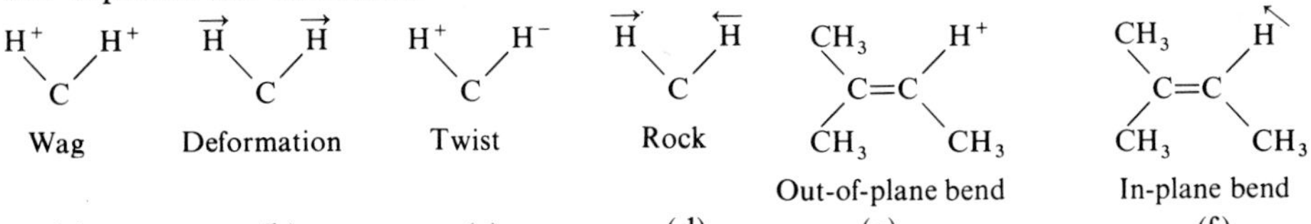

Wag (a) Deformation (b) Twist (c) Rock (d) Out-of-plane bend (e) In-plane bend (f)

A49

out-of-plane — in-plane

A50 (b) and (d) are incorrect. The captions are in fact interchanged.

Infrared Absorption Spectra

F51 The vibrations we have been considering from **F39** onwards are typical of most vibrations in organic molecules. Each one will have its own frequency and will give rise to absorption of infrared radiation of the same frequency. If, for a particular molecule, we construct a diagram representing absorption of infrared radiation versus its frequency, the result would consist of a series of **absorption bands** known as a **spectrum** e.g. Fig. 1.7. The form of the 'spectrum' shown in Fig. 1.7 corresponds to the conventional way of drawing a graph with the ordinate

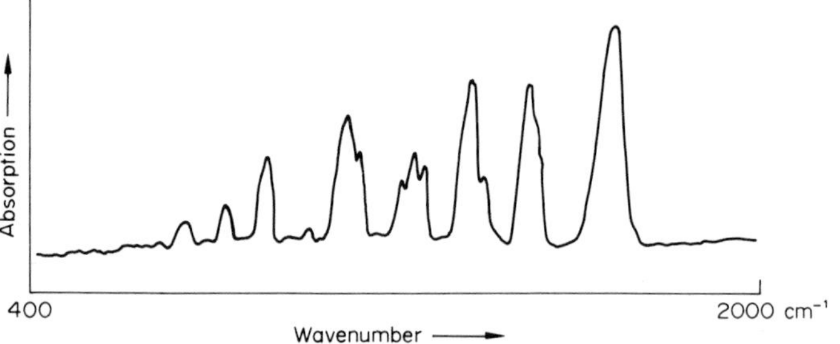

Fig. 1.7. Part of the *infrared spectrum* of cyclohexanone.

and abscissa scales increasing from bottom to top and left to right respectively. The majority of current infrared spectrophotometers however present the spectrum with the wavenumber scale increasing from **right to left.** Furthermore the ordinate scale is normally calibrated in '% transmission' with 100% transmission (0% absorption) at the top of the chart. As a result, spectra are usually presented both by the instrument and in publications inverted and reversed from left to right as compared with Fig. 1.7. The conventional presentation of the spectrum in Fig. 1.7 is shown in Fig. 1.8.

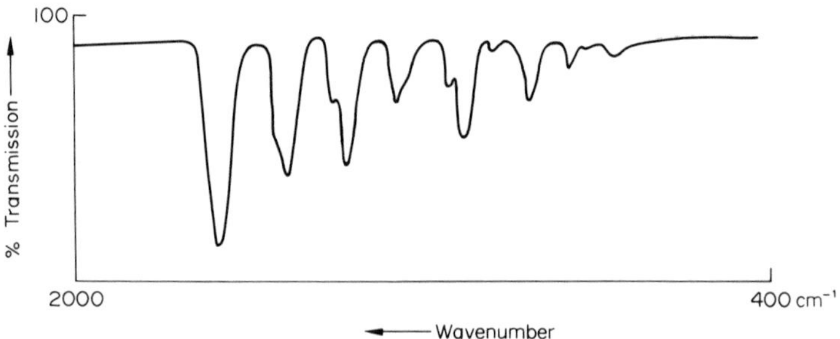

Fig. 1.8

State whether each of the following statements is true or false.

(a) An infrared spectrum is a plot of absorption versus frequency.
(b) An infrared spectrum consists of a series of absorption bands, each one representing absorption of radiation by a stretching vibration.
(c) The frequency at which an absorption band occurs in a spectrum is equal to the frequency of the vibration associated with it.

F52 Each band in an infrared spectrum corresponds to the promotion of a molecule from the vibrational ground state to a vibrational excited state by absorption of infrared radiation of frequency ν (Fig. 1.9). You recall we said earlier that vibrational levels in a molecule are subdivided into closely spaced rotational levels, so that the precise energy of absorption will depend on the particular rotational levels involved (Fig. 1.10). That is to say, transitions between the ground state and the first excited vibrational state are associated with the absorption of infrared radiation, not of a unique frequency (Fig. 1.11), but of a whole series of closely spaced frequencies (Fig. 1.12). This type of spectrum is indeed observed for gas samples if a high performance spectrophotometer is used.

Match each of the groups of absorptions (a), (b) and (c) (Fig. 1.12) with one of the groups of transitions (p), (q) and (r) (Fig. 1.10).

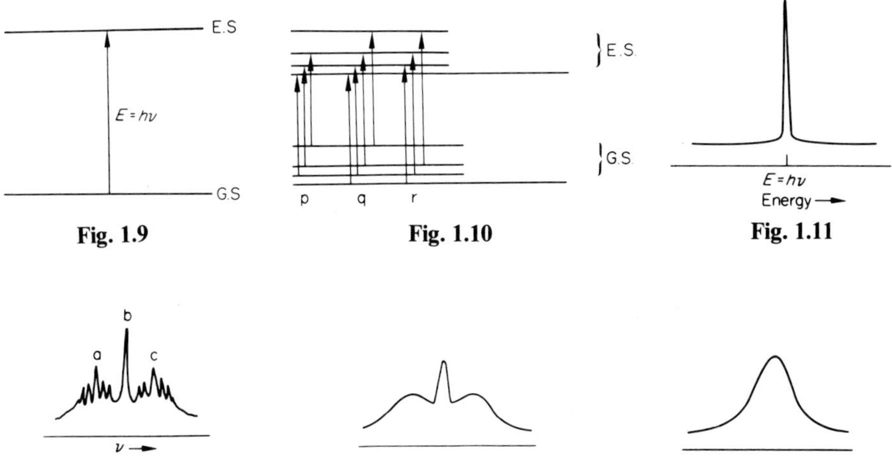

Fig. 1.9

Fig. 1.10

Fig. 1.11

Fig. 1.12. Gas sample, high resolution.

Fig. 1.13. Gas sample, low resolution.

Fig. 1.14. Liquid sample.

A51 (a) true (unless one insists on the use of the term wavenumber instead of frequency)
(b) false (absorption bands corresponding to bending vibrations will also occur)
(c) true (in relation to information you have so far).

F53 The type of spectrum obtained in Fig. 1.12 can only be obtained with sophisticated instruments capable of separating very closely spaced bands. With the type of instrument used for routine infrared spectroscopy the spectrum appears as shown in Fig. 1.13. Here resolution is lost because of instrumental factors. With liquids and solids, however, the fine structure is lost completely because of interactions between molecules in the condensed phase. This results in a band of the general shape shown in Fig. 1.14. Will the spectrum of a solution on a sophisticated instrument contain bands of the type shown in Fig. 1.12, Fig. 1.13 or Fig. 1.14?

- - -

A52 (a) with (p); (b) with (q); (c) with (r).

The horizontal scale in Fig. 1.12 is one of frequency and this is directly proportional to the energy of the corresponding transition. The energies of the transitions shown in Fig. 1.10 increase in the order $p < q < r$. The frequencies of the groups of absorptions in Fig. 1.12 are in the order (a) < (b) < (c).

F54 Even in a relatively simple molecule such as cyclohexanone

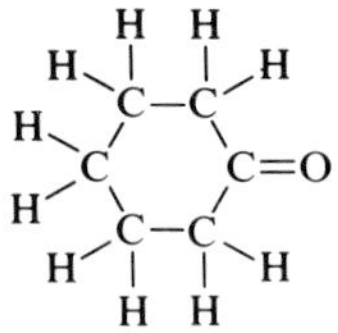

```
      H  H
    H\|  |/H
      C—C
  H\ /    \
    C      C=O
  H/ \    /
      C—C
    H/|  |\H
      H  H
```

it is clear that a very large number of vibrations is possible. In addition to the C=O stretch and the many C—H stretches and bends there are also C—C stretches and many vibrations involving bending of the C—C—C and C—C—H angles. If each vibration was equally capable of absorbing infrared radiation (and each had a different frequency) we should have a very complex state of affairs.

List the types of vibration which may occur in the allyl bromide molecule $CH_2{=}CHCH_2Br$.

- - -

A53 Figure 1.14. Since the fine structure is lost because of the state of the sample and not instrumental factors, the performance of the instrument used will not affect the spectrum.

F55 In practice the potential complexity of the infrared spectra of large molecules is simplified by three factors:

(i) The frequencies of some vibrations will lie outside the range of infrared radiation normally used in recording the spectrum.
(ii) By no means all vibrations give rise to strong absorption of infrared radiation.
(iii) Some bands are so close to each other that they coalesce. This is particularly true where we have a number of very similar groups in the molecule.

Consider the first factor.

You should by now be aware of the normal range of radiation generally employed in infrared spectroscopy. Quote this range in wavenumbers.

A54 C=C stretch   C—H stretch   C—C stretch   C—Br stretch
In-plane and out-of-plane bend of the terminal CH_2 group.
In-plane and out-of-plane bend of the =CH group.
Deformation vibrations within the CH_2Br group. Rocks and wags of the CH_2Br methylene group. Twisting vibrations about the C=C and the C—C bonds.
Bends involving the C—C—C skeleton.

F56 While there are no fundamental vibrations which give rise to absorption of infrared radiation beyond the upper limit, most molecules possess vibrations of low frequency which absorb below 400 cm^{-1}. These include some of the bending frequencies of carbon chains, bending vibrations involving heavy atoms such as halogens and most twisting vibrations.

On the basis of the last sentence pick out **three** vibrations from the following list which may absorb below 400 cm^{-1}.

In cyclohexanone:

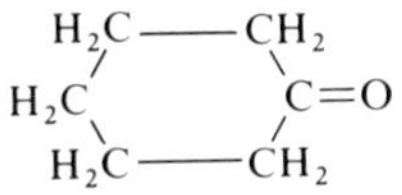

```
   H2C ——— CH2
   /          \
H2C            C=O
   \          /
   H2C ——— CH2
```

(a) C=O stretch
(b) CH_2 wag
(c) a ring bending vibration

In allyl bromide:

CH_2=CH—CH_2—Br

(d) C—Br stretch
(e) CC Br bend
(f) twisting of the CH_2Br group about the C—C bond connecting it to the vinyl group

---

A55 400—5000 cm^{-1}. Incidentally the bottom end of this range is not arbitrary but is determined by instrumental factors. Until recently most infrared spectrophotometers were fitted with sodium chloride prisms which do not transmit infrared radiation below 650 cm^{-1}. Some of these had interchangeable potassium bromide prisms which extended the range down to 400 cm^{-1}. This extended range to 400 cm^{-1} is what we refer to as the 'normal' lower limit. Many modern instruments are fitted with gratings instead of prisms but in most cases these work to the same lower limit. More sophisticated (expensive) instruments operating below 400 cm^{-1} are available but they are not generally used for routine analytical work. (This information is for your interest only and is not essential to the understanding of the material to follow.)

F57 Now let us consider the second factor which simplifies infrared spectra, that is, the fact that not all vibrations give rise to strong absorption of infrared radiation. We have already seen that infrared radiation can only be absorbed by a molecule if its energy, $h\nu$, exactly matches the energy difference between two energy levels of the molecule. In practice, this means that the frequency of the radiation must be the same as the frequency of one of the normal modes of vibration of the molecule. Not all vibrations, however, are associated with the absorption of infrared radiation and the operation of **selection rules** enables us to pick out those that are.

The operation of a selection rule is likely to *increase/decrease* the number of vibrations giving rise to infrared absorption bands.

- -

A56 (c), (e), (f).

F58 Is the following statement true or false?

All transitions between available vibrational energy levels in a molecule can be induced by interaction with infrared radiation.

- -

A57 decrease.

F59 The basic selection rule for i.r. spectroscopy may be expressed in the following terms.

'In order for a vibration to give rise to absorption of infrared radiation, it must cause a change of dipole moment in the molecule'. Therefore a strong absorption band in an infrared spectrum implies that during the vibration concerned, there is a *large change/no change* in the dipole moment of the molecule.

- -

A58 false. See **F57**.

F60 The stretching vibration of the carbonyl group involves changes in the distance between the carbon and oxygen atoms. Now we know that because of the difference in the electronegativities of carbon and oxygen the carbonyl group is polarized in the following way.

$$\overset{\delta+}{\mathrm{C}}=\overset{\delta-}{\mathrm{O}}$$

Furthermore, as the bond dipole is a function of the distance between the charges, any change in the C=O bond length alters the dipole moment. Thus, we expect the carbonyl group stretching vibration to be strongly absorbing in the infrared.

In carbon dioxide, however, which has effectively two carbonyl groups, two types of stretching vibration may occur.

$\overset{\leftarrow}{O}=C=\vec{O}$ symmetric

$\vec{O}=\overset{\leftarrow}{C}=\vec{O}$ antisymmetric

One of these is not observed in the infrared spectrum of carbon dioxide.

Which one?

A59 large change.

F61 In contrast with the carbonyl group the double bond of ethylene is electrically symmetrical, i.e. there is no charge separation. Consequently there is *some change/no change* in dipole moment during a stretching vibration, and such a vibration *does/does not* absorb infrared radiation and *is not/is* observed in the infrared spectrum of the compound.

A69 symmetric. Carbon dioxide has zero dipole moment in its equilibrium position since both C=O bonds are of equal length and the bond dipoles associated with each are exactly balanced. During the **symmetric** stretching vibration, the bond lengths are equal at all times, and the dipole moment therefore remains unchanged throughout. During the **antisymmetric** stretching vibration, however, one bond is lengthened while the other is shortened. The bond dipoles do not balance except when the molecule passes through the equilibrium position.

F62 In practice, the difference between absorption bands allowed or forbidden by the selection rule is often not as clear cut as we have implied in the examples given so far. The change in dipole moment during vibration may be very small, with the result that the absorption may be very weak or even unobservable. For example, propylene is unsymmetrical and we expect a small separation of charge.

$\overbrace{CH_3-CH}^{\delta+}=\overset{\delta-}{CH_2}$

Consequently we would expect the C=C stretching vibration of propylene to be observed as a *very weak/strong* absorption in the infrared spectrum of the compound.

A61 no change, does not, is not.

F63 Would you expect absorption by the double bond stretching vibration of 1,1′-difluoroethylene to be stronger or weaker than that of propylene?

A62 very weak. Since it is often difficult to estimate the magnitude of the change in dipole moment during a vibration (particularly in the case of bending modes), it is not always possible to predict whether a given vibration will give rise to a strong or weak band in the spectrum.

F64 Which of the following isomeric alkenes has the strongest absorption band arising from the C=C stretching vibration?

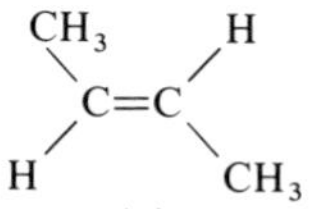

$CH_3(H)C{=}C(H)CH_3$ (a) $\qquad (CH_3)_2C{=}CH_2$ (b)

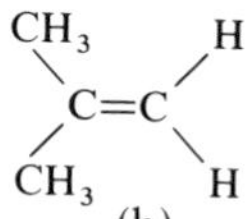

A63 stronger. Fluorine is strongly electron withdrawing and the double bond in this molecule would be heavily polarized.

F65 The dipole moment of the molecule depends not only on the magnitude of the bond dipoles but also on the direction in which they act. In 'pure' bending vibrations therefore (where there is no change in bond length), changes in dipole moment depend upon the changes in direction of the bonds relative to each other. State whether the following vibrations will give rise to absorption in the infrared region.

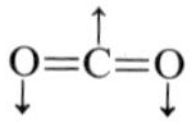

O=C=O (a) $\qquad$ H—C≡C—H (b) $\qquad$ $H_2C{=}CH_2$ (c)

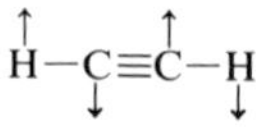

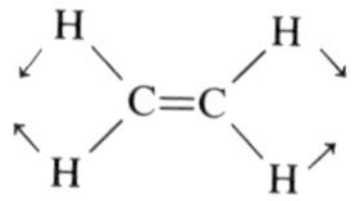

A64 (b). In (a) the inductive effects of the two methyl groups are exactly balanced because (a) has a centre of symmetry. In fact the C=C stretching vibration in (a) does not give rise to the absorption of infrared radiation—it is described as 'infrared inactive'.

F66 It is more difficult to predict the intensity of absorptions due to bending vibrations than those due to stretching vibrations. One case, which is clearer than most, is the out-of-plane bending vibration in vinyl compounds. Consider the C—H bond in trichloroethylene.

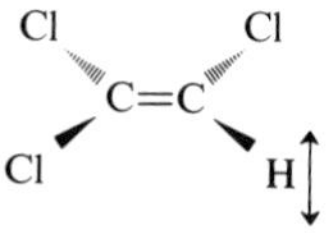

$Cl_2C{=}CClH$

The out-of-plane vibration involving this bond (as shown above) causes a change in dipole moment. In which of the following directions does it occur:

(a) in the plane of the paper along the C=C bond;
(b) parallel to the plane of the paper, perpendicular to the C=C bond;
(c) perpendicular to the plane of the paper?

A65 (a) is infrared active. (b) and (c) are inactive. Consider the shapes of the molecules at the extremes of their vibrations.

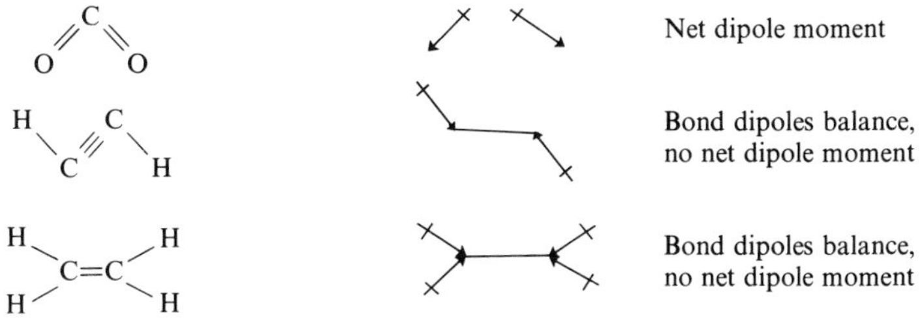

Net dipole moment

Bond dipoles balance, no net dipole moment

Bond dipoles balance, no net dipole moment

All these molecules have no dipole moment in the equilibrium position and it is only in the case of (a) that a dipole moment is created by the deformation.

F67 *cis*-Dichloroethylene can have two C—H out-of-plane bending vibrations.

(a) (b)

Which of the two would you expect to absorb most strongly in the infrared?

A66 (b). Notice here it is not a bond dipole which changes but the direction in which the dipole acts.

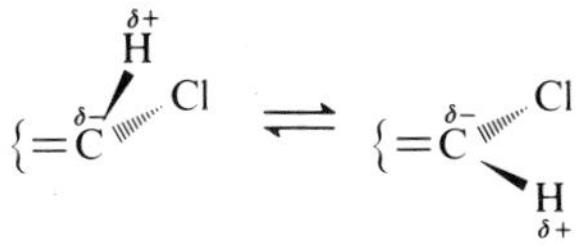

This affects the dipole moment of the molecule.

F68 You will notice that the symmetry of the vibrations concerned and of the molecule as a whole plays an important part in determining whether a particular absorption band is strong or weak. Which of the following generalizations are true?

(a) Bands corresponding to symmetric vibrations are generally weaker than the corresponding antisymmetric vibrations.
(b) A very symmetrical molecule gives rise to a much more complicated spectrum than does an unsymmetrical molecule of similar size.

(c) Even when a molecule has little or no symmetry, as in the case of the majority of organic compounds, the stretching vibrations involving bonds between different atoms (e.g. C—H) usually give rise to stronger absorption than those corresponding to stretching vibrations involving similar atoms (e.g. C—C, C=C, C≡C, N=N).

A67 (a) In this vibration the effects of the motions of the two hydrogen atoms are additive. Vibration (b) is more symmetric and leads to weaker absorptions.

F69 Finally we come to the third factor which simplifies spectra (**F55**), the fact that vibrations of a similar type have similar frequencies. A molecule such as chloroform shows a single absorption band near 3000 cm^{-1} which is associated with the stretching of the one C—H bond in this molecule. In methylene chloride two C—H stretching frequencies occur in the same region associated with the two C—H bonds in the molecule. In methyl chloride however, although there are three C—H bonds, only two C—H stretching frequencies appear in the spectrum. The third vibration happens to have the same frequency as one of the others.

In cyclohexanone, in spite of the fact that there are 10 C—H bonds only two bands appear in the C—H stretching region of the spectrum. This is because the vibration of any one of the CH groups may be considered independent of the others. The two bands originating from the group therefore have almost exactly the same frequencies as those of the others.

In a similar way only two C—H stretching bands are observed in molecules containing more than one methyl group and furthermore these have practically the same frequencies as the two CH_2 bands. How many absorption bands would you expect to observe in the C—H stretching region of the infrared spectrum of each of the following compounds?

(a) $(CH_3)_3CCH_2CH_2CH_3$

(b) $CH_3COCH_2OCH_3$

(c) CH_3N ring: $CH_2—CH_2$ / $CH_2—CH_2$ (N bonded to both CH_2 groups, which are joined to each other)

A68 (a) and (c) are true. (b) is false because symmetric molecules will have some vibrations which involve little or no change in dipole moment. Vibrations of unsymmetric molecules nearly always involve a change in dipole moment.

F70 Although the illustrations used in **F69** involve the C—H stretching vibrations of saturated aliphatic compounds, the same simplifying factor is equally applicable to the stretching vibrations of most functional groups present in organic molecules.

The same is also true of some CH bending vibrations. A convincing demonstration of this effect is illustrated in Fig. 1.15 which compares the spectra of two molecules having the same functional groups but differing widely in size.

Select a simple molecule from list A (p. 30) which is likely to have a similar infrared spectrum to each complicated molecule in list B.

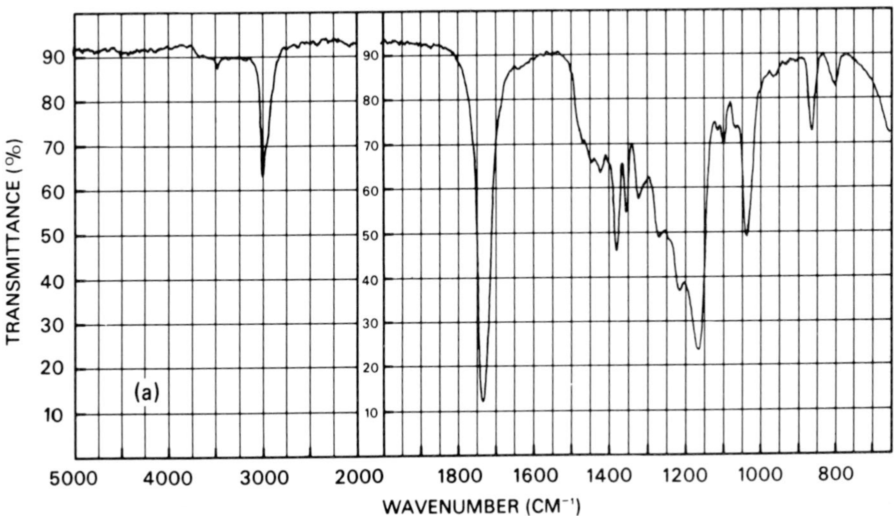

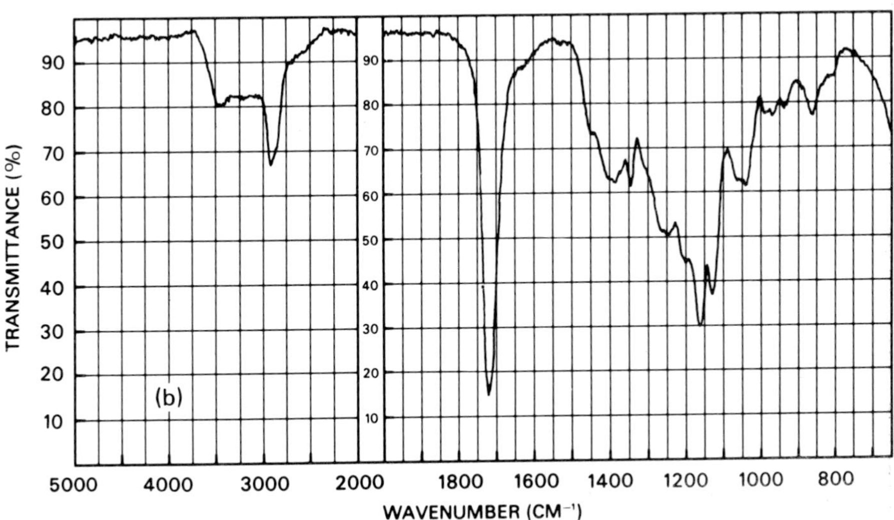

Fig. 1.15. **(a)** Spectrum of diethyl succinate $C_2H_5OCOCH_2CH_2COOC_2H_5$. **(b)** Spectrum of polydiethyleneglycol succinate $(-CH_2CH_2OCOCH_2CH_2COO-)_n$.

List A

(a) OCHCH(OH)CH_2OH

(b) NH_2COCH_2COOH

(c)

List B

(a)

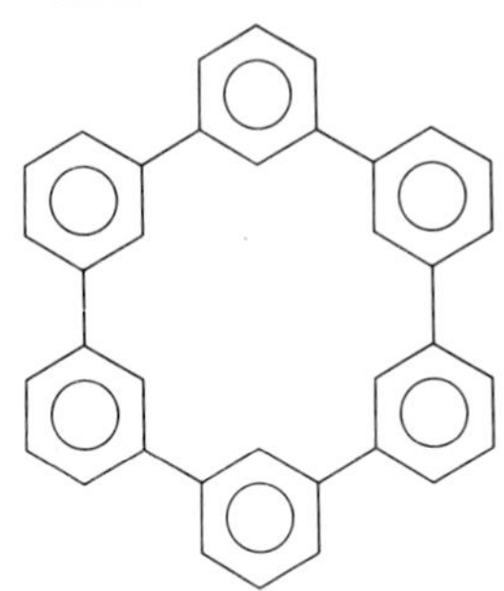

(b) $NH_2CH_2CONHCH_2CONHCH_2CONHCH_2$
HOOCCHNHCO

(c) CHOH—CHOH
CHOH CHOH
CHO CH_2OH

A69 You might expect 4 in each case, but in fact you would normally observe only 2. Each of the two bands of the CH_3 group is very close to one of the bands of the CH_2 group and the type of spectrometer used for routine work is often unable to separate the two pairs. More sophisticated instruments however, are able to show greater detail and all four bands can be observed.

$—CH_3$ 2962 and 2872 cm^{-1} $—CH_2—$ 2926 and 2853 cm^{-1}.

F71 Summarize the three main factors which help to simplify infrared spectra.

A70 *List A* *List B*

List A		List B
(a)	with	(c)
(b)	with	(b)
(c)	with	(a)

F72 State whether the following statements are true or false:

(a) Under normal circumstances the C—H stretching region of the spectrum of a saturated aliphatic compound consists of two bands.

(b) The double bond stretching vibration in cyclohexene is expected to be strongly absorbing in the infrared.

A71 The frequencies of some vibrations will lie outside the range of infrared radiation which is normally used in recording spectra.

Many vibrations do not give rise to strong absorption of infrared radiation.

Many absorption bands are so close to one another that they coalesce.

F73 Consider the vibrational properties of *trans*-2,3-diiodo-2-butene.

$$\begin{array}{ccc} CH_3 & & I \\ & C{=}C & \\ I & & CH_3 \end{array}$$

Illustrate each of the three simplifying factors discussed above by three vibrations of this molecule, one for each factor.

A72 (a) true (b) false. The stretching of the C=C bond is a symmetrical vibration in this molecule.

A73

Bends involving iodine All twists	beyond the range of normal spectrometers
All stretches and bends which are symmetrical about the centre of the C=C bond	infrared inactive (no change in dipole moment)

The vibrations within one methyl group will coalesce with those of the other.

Overtones and Combination Bands

F74 Until now, we have been discussing the principal features of the infrared spectrum of a compound in terms of its vibrations. In addition to these, however, there are very weak secondary features due to overtones and combinations of vibrations. Overtones can be illustrated by analogy with the vibrations of a stretched string and by drawing the transitions involved on an energy level diagram.

stretched string vibration:

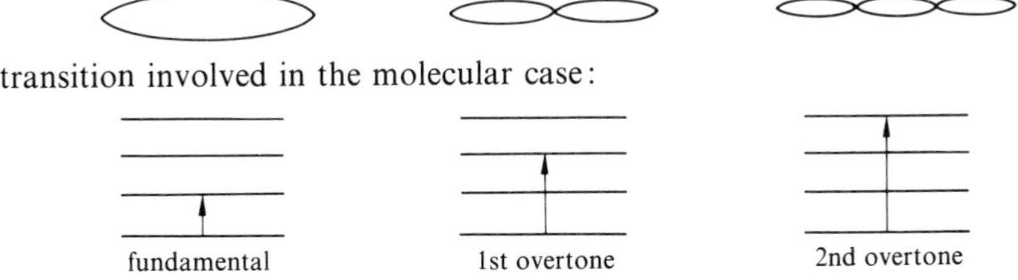

Note that the wavelength of the 1st overtone is exactly half that of the fundamental. Its frequency is therefore twice that of the fundamental. If the fundamental vibration has a wavenumber of 1000 cm^{-1} and the energy levels are equally spaced, what are the wavenumbers of the first and second overtones?

F75 Combination bands occur when two fundamental vibrations of frequency ν_1 and ν_2 are simultaneously excited by the absorption of infrared radiation of frequency $(\nu_1 + \nu_2)$ giving rise to a band in the spectrum at this frequency.

Suppose a molecule had three fundamental vibrations at 800, 2000 and 2500 cm^{-1}. Write down the wavenumbers of

(a) the three first overtones
(b) the combination bands of the type $(\nu_1 + \nu_2)$.

A74 1st overtone at 2000 cm^{-1}, 2nd overtone at 3000 cm^{-1}.

F76 Overtones and combinations are invariably very much weaker than the fundamentals from which they are derived and in most cases they are weaker than all the other fundamentals observed in the spectrum. Consequently, in the majority of cases the overtone or combination will be easily recognised as such and not confused with a fundamental. The infrared spectrum of sulphur dioxide is given as the table of wavenumbers below, in which a qualitative indication of the intensity of each band is also shown. Select the three fundamental bands from this information and identify the other bands as overtones or combinations giving the fundamentals concerned in each case.

519	very strong
1151	very strong
1361	strong
1871	very weak
2305	very weak
2499	moderate

A75 First overtones 1600, 4000, 5000 cm^{-1}
Combinations 2800, 3300, 4500 cm^{-1}

In practice it is found that the spacing between energy levels for a given vibration is not equal but that the levels slowly converge. Consequently the first overtone is usually found at a frequency slightly lower than twice the fundamental frequency. This effect is very slight however and the approximation of equally spaced levels is good enough for most purposes.

F77 Although overtones and combinations are normally extremely weak, occasionally they do appear with anomalously high intensities, comparable with weak fundamentals. This occurs through a phenomenon known as 'Fermi resonance' where the overtone or combination has approximately the same frequency as another fundamental. The effect is greatest when the resonance is exact, that is when overtone and fundamental have exactly the same frequency. In this

Fig. 1.16. (a) Condition for Fermi resonance. **(b)** Result of Fermi resonance.

case two bands of equal intensity fairly close together and on either side of the resonance frequency are observed—a so-called 'Fermi doublet'.

An example of Fermi resonance is found in the spectrum of salicylaldehyde

OH

CHO

This molecule has one non-aromatic C—H bond (in the CHO group), but in the region where the band corresponding to the stretching of this bond is expected two bands of approximately equal intensity are found at 2732 and 2833 cm^{-1}. This is explained by Fermi resonance between the C—H stretching vibration of the —CHO group and an overtone of a CH bending mode, thus $2 \times 1390 = 2780\ cm^{-1}$.

Carbon tetrachloride has four fundamental frequencies. Three of these occur at 217, 313 and 459 cm^{-1}. The fourth frequency is expected to occur in the 700—800 cm^{-1} region. The infrared spectrum in this region shows a doublet of almost equal intensity at 762 and 791 cm^{-1}. How do you account for the apparent observation of five fundamentals in this molecule?

A76 Fundamentals 519 1151 1361

$1871 \approx 519 + 1361$ (combinations)
$2499 \approx 1151 + 1361$ (combinations)
$2305 \approx 2 \times 1151$ (overtone)

F78 Two other conditions, apart from coincidence of frequency, must also be met in order that Fermi resonance may occur.

(a) The overtone and fundamental must have the same symmetry properties.
(b) The overtone and fundamental must be predominantly associated with a group of atoms in the same part of the molecule.

(b) is of particular importance in large organic molecules having low overall symmetry. How is condition (b) fulfilled in the case of salicylaldehyde? Is the bending mode whose overtone is in Fermi resonance with the CH stretch more likely to be an in-plane or out-of-plane vibration (point (a) above)?

A77 Fermi resonance between the expected fourth fundamental and the combination $459 + 313 = 772$. Note that when Fermi resonance occurs it is impossible to say which member of the doublet is the fundamental and which the overtone or combination. If the numerical value of the fundamental frequency is required for a calculation it is usual to take the mean frequency of the two components.

F79 How do the intensity and frequency of a band assist in its recognition as (a) an overtone and (b) a combination?

A78 All the vibrations concerned occur within the —CHO group. In-plane, since the C—H stretching must be an in-plane vibration.

F80 If all the fundamental vibrations of a molecule occur in the range 500—1000 cm^{-1} one would expect the first overtones to be *<500 cm^{-1}/in the range 500—1000 cm^{-1}/above 1000 cm^{-1}/above 4000 cm^{-1}*.

A79 (a) An overtone has low intensity and occurs at a frequency which is twice or possibly three times the frequency of a strong fundamental.
(b) A combination has low intensity and occurs at a frequency which is the sum of 2 (or possibly 3) fundamental frequencies.

F81 Label the following remarks as referring to an overtone band, a combination band or Fermi resonance.

(i) 'Carbonyl stretching bands sometimes appear as multiple absorptions even when there is only one carbonyl group in the molecule'.
(ii) 'The precise position of the weak band near 1900 cm^{-1} in this series of compounds is dependent on the frequencies of two other vibrations . . . '

A80 above 1000 cm^{-1}.

A81 (i) Fermi resonance. (ii) Combination band. The multiple absorption includes only one fundamental.

Revision Summary

Parameters associated with electromagnetic radiation

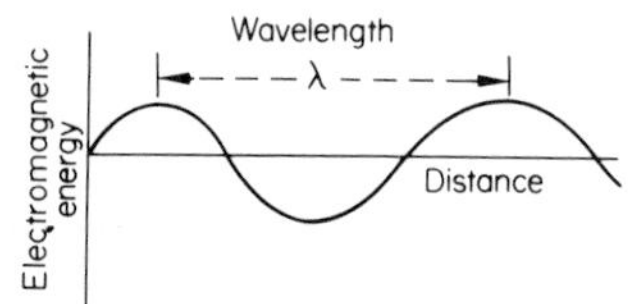

Number of waves in 1 cm = wavenumber
$\bar{\nu}$, cm^{-1}

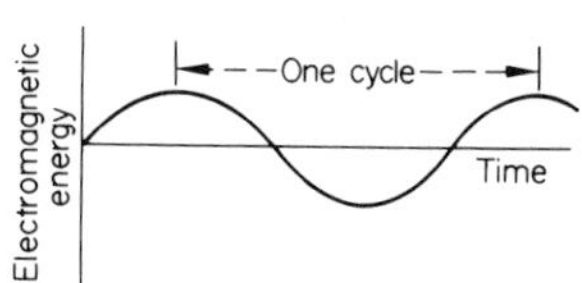

Number of cycles in 1 s = frequency
ν, cycles/second (Hertz)

Electromagnetic radiation of all wavelengths travels at approximately 3×10^{10} cm s^{-1} in a vacuum. This constant is given the symbol c

$$c = \lambda\nu \qquad \bar{\nu} = \frac{1}{\lambda}$$

The wavelength range of infrared radiation for routine spectroscopic use is 2—25 micrometres (1 micrometre (μm) = 10^{-4} cm). This range is equivalent to

5000—400 cm^{-1}. Energy of electromagnetic radiation, $E = h\nu$, h = Planck's constant, $\therefore E = hc/\lambda = hc\bar{\nu}$, i.e. $E \propto 1/\lambda$; $E \propto \bar{\nu}$.

Interaction of radiation with matter

Matter absorbs infrared radiation in quantities equivalent to energy differences between molecular vibrational levels, individual molecules undergoing **transitions** from **ground** to **excited** states.

Stretching vibrations, e.g. $\overleftarrow{\mathrm{H}}-\overrightarrow{\mathrm{Cl}}$ approximate to those of a stretched spring and obey Hooke's Law,

$$\omega = \frac{1}{2\pi}\sqrt{\frac{k}{m_1 m_2/(m_1 + m_2)}}$$

where ω is the frequency of vibration, k is the force constant of the bond (directly related to bond strength) and m_1 and m_2 are the masses of the two atoms. For absorption, $\omega = \nu$.

Vibration of molecules

Bending vibrations involve changes in bond angle. Stretching vibrations involve changes in bond length.

Examples:

Antisym. Sym. — C—H stretch

In-plane (deformation), Out-of-plane (wag), Out-of-plane (twist), In-plane (rock) — CH bend

C—C C=O O—H — Stretching vibrations

$CH_3(CH_3)C{=}C(H\pm)(CH_3)$ C—H out-of-plane bend

Spectra

An infrared spectrum is a plot of absorption versus 'frequency' of radiation, and consists of a series of **absorption bands**. The number of molecular vibrations which can give rise to individual i.r. absorption bands is limited by three factors:

(i) Frequencies of some vibrations fall outside the range of a routine spectrophotometer (e.g. bending vibrations involving halogens, some bending vibrations of carbon chains, most twisting vibrations).

(ii) Many vibrations do not absorb infrared radiation because they do not fulfil the requirement of a change in dipole moment during the vibration. Many 'symmetrical' vibrations fall in this category e.g. the C=C stretch in ethylene.

(iii) Vibrations involving similar groups within a molecule are likely to absorb radiation of the same frequency. For example, all the CH_2 stretching vibrations of cyclohexanone give rise to only two observable absorptions.

There are three types of absorption band which may occur in a spectrum but which do not arise directly from a fundamental vibration. They are:

overtone bands
combination bands
bands due to Fermi resonance.

Now have another go at the multiple choice test at the beginning of this Part. Write your answers in a vertical column and mark the script as before. Then consult the remarks below.

0—2 Check that you have followed the instructions at the beginning of the book.

3—6 Consult your instructor as to whether you should go on or seek extra tuition on basic principles of the subject. If this is inappropriate, re-read the frames pertinent to those questions which you answered incorrectly and, when you are satisfied you fully understand the material, read the paragraph at the end of this Part and start Part 2.

More than 6 Revise the material pertinent to the questions you answered incorrectly. If your instructor does not want you to attempt the questions or consult the bibliography which follows, read the paragraph at the end of this part and continue with Part 2.

Questions

1. Describe from first principles the meaning of the statement:

   'The carbonyl group of acetone absorbs at 1720 cm^{-1} in the infrared'.

2. Draw diagrams of the following vibrations:

   methyl twist; in-plane bend of the alkene carbon–hydrogen bond in trichlorethylene; an aryl–hydrogen bond in-plane bend of toluene; the symmetric NH stretch of methylamine; the symmetric C—Cl stretch of carbon tetrachloride; the out-of-plane C—H bend of acetaldehyde; the methylene rocking vibration of dichloromethane.

3. Describe fully the three principal factors which help to simplify infrared spectra, and any factors which may give rise to more bands than one would expect from fundamental vibrational modes alone.

Further Reading

J. R. Dyer, *Applications of Absorption Spectroscopy of Organic Compounds*, Prentice Hall, Englewood Cliffs, N.J., 1965, Chapter 1 and introduction to Chapter 3.
J. C. Brand and G. Eglinton, *Applications of Spectroscopy to Organic Chemistry, Oldbourne*, London, 1965.
A. Liberles, *Introduction to Theoretical Organic Chemistry*, Macmillan, London, 1968, p. 241.
C. N. Banwell, *Fundamentals of Molecular Spectroscopy*, McGraw-Hill, New York, 2nd Edn., 1972, Chapter 1 and introduction to Chapter 3.
D. H. Whiffen, *Spectroscopy*, Longmans, London, 1966, Chapters 1, 2 and 8.

R. M. Silverstein and G. C. Bassler, *Spectrometric Identification of Organic Compounds*, John Wiley and Sons, New York, 1967, Chapter 3.
J. C. P. Scharz, *Physical Methods in Organic Compounds*, Oliver and Boyd, Edinburgh, 1964, Chapter 1 and introductory part to Chapter 3.
E. F. H. Brittain, W. O. George and C. H. J. Wells, *Introduction to Molecular Spectroscopy*, Academic Press, London, 1970, Chapter 1 and Chapter 3.
Also introductory sections of the following:
K. Nakanishi, *Infrared Absorption Spectroscopy*, Holden Day, New York, 1962.
C. N. Rao, *Chemical Applications of Infrared Spectroscopy*, Academic Press, New York, 1964.
Mansel Davis, *Infrared Spectroscopy and Molecular Structure*, Elsevier, Amsterdam, 1963.
R. T. Conley, *Infrared Spectroscopy*, Allyn and Bacon, Boston, 1966.
A. D. Cross and R. A. Jones, *An Introduction to Practical Infrared Spectroscopy*, Butterworths, London, 1969.
M. Avram and Gh. D. Mateescu, *Infrared Spectroscopy, Applications in Organic Chemistry*, Wiley-Interscience, New York, 1972.
J. H. van der Maas, *Basic Infrared Spectroscopy*, 2nd Edn. Heyden and Son, London, 1972.
S. F. Dyke, A. J. Floyd, M. Sainsbury and R. S. Theobald, *Organic Spectroscopy—an Introduction*, Penguin, London, 1971.
F. Scheinmann, *Introduction to Spectroscopic Methods for the Identification of Organic Compounds*, Volume 1, Pergamon Press, Oxford, 1970.

Having mastered the fundamental principles behind infrared spectroscopy, and in particular, the fact that small groups within a molecule can give rise to absorption bands at particular frequencies, you are now in a position to study the factors which control the precise value of group vibrational frequencies and the use of empirical data in the interpretation of infrared spectra. Before doing this, however, you should have some knowledge of the experimental techniques involved in infrared spectroscopy since they impose some limitations on its application. This is the subject of Part 2.

PART 2
EXPERIMENTAL CONSIDERATIONS

Aim

The aim of this part is to acquaint the student with the practical scope and limitations of the use of infrared spectroscopy as a tool in organic structural analysis. He will acquire sufficient knowledge of the operation of infrared spectrophotometers and relative merits of different sample preparation techniques to be able to recognize situations where experimental factors can influence the interpretation of an infrared spectrum.

New Terms and Concepts

Spectrophotometer, monochromator, source, detector, recorder, double beam instrument, optical null, optical wedge, slit programme, double beam instrument, Nujol mull, attentuated total reflectance.

Prior Knowledge

Part 1, basic optical function of a lens and a mirror, diffraction grating, prism, the difference between direct and alternating current, critical angle for refraction, refractive index. Meaning of the terms gas liquid chromatography and molarity.

Objectives for Part 2

When you have completed Part 2 you should be able to:

1. Describe briefly the function of the

   (a) source
   (b) monochromator
   (c) detector
   (d) recorder

   in an infrared spectrophotometer.

2. Recognize a correct sequence of events within a spectrophotometer which leads to a recorded spectrum.

3. Identify situations where the sample whose spectrum is required is best prepared as a

   (a) solution
   (b) thin liquid film
   (c) smooth paste (mull)
   (d) pressed disc with an alkali halide
   (e) gas or vapour
   (f) surface from which the i.r. beam is reflected

   and to recognize the implications of each method pertinent to the interpretation of the spectrum obtained.

4. Identify structural features in a spectrum whose interpretation may depend on the type of sample preparation used.

PART 2: EXPERIMENTAL CONSIDERATIONS

Introduction

Although in most instances the interpretation of an infrared spectrum is only marginally dependent on the experimental conditions under which it was recorded, occasions can arise when practical considerations are very important. This section of the programme constitutes a brief summary of pertinent instrumental and sampling techniques, which will enable you to place the theoretical subject matter in the context of the laboratory. The material is not programmed, but there are some questions and suggestions for further reading on p. 56.

The Infrared Spectrophotometer

It was explained in Part 1 that the infrared spectrum of a substance is a graph which shows how the absorbance changes with wavelength (or wavenumber). The instrument used to obtain this spectrum is an infrared spectrophotometer. Its operation is based on the principle that continuous infrared radiation (i.e. radiation whose intensity versus wavelength curve changes very slowly across the infrared region) is passed through a sample and the transmitted radiation is analysed by a monochromator and passed to a detector. The system registers the extent to which the sample absorbs the radiation as the wavelength is changed and indicates the wavelengths or more commonly, the wavenumbers at which the absorbance passes through a maximum. These maxima are the absorption bands of the sample.

The layout of a typical infrared spectrophotometer is shown as a block diagram in Fig. 2.1.

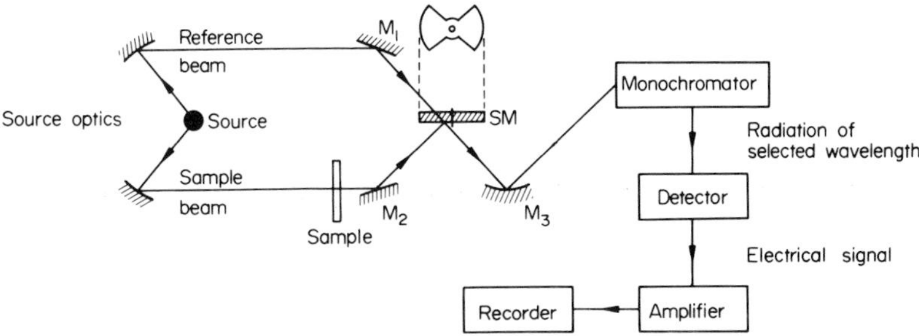

Fig. 2.1. Schematic diagram of a double-beam infrared spectrophotometer.

The source consists of a rod or tube containing rare-earth oxides which is heated to 1200—1500°C (bright 'orange heat') either directly by passing a current through it (as in the Nernst filament) or indirectly by means of a nichrome winding inside the tube.

Radiation from the source is split into two beams by the mirrors comprising the 'source optics' which bring both beams to a focus in the sampling area, the sample being placed at the focus of the sample beam. The sample and reference beams are then directed on to the entrance slit of the monochromator by means of the mirror system $M_{1,2,3}$ and a rotating sector mirror, SM. The sector mirror is driven by a motor at a speed such that each beam is selected alternately twenty times a second. (The actual chopping frequency varies from one instrument to another.) Glass absorbs strongly in the infrared region so all the mirrors in the optical system (and in the monochromator) must have the reflecting material coated on the front surface. Similarly concave mirrors are used for focussing the beams since any lens system would have to be made out of rock-salt, an expensive material in optics.

The function of **the monochromator** is to select a particular wavelength (actually a narrow band of wavelengths) which is passed to the detector. The selection is made continuously so that the instrument scans from one end of its range to the other. This is achieved by rotating a prism or diffraction grating. Prism monochromators fitted with rock-salt (NaCl) prisms have been extensively used for infrared work and these are quite satisfactory for routine purposes. Their range in terms of wavenumbers is limited to 650—5000 cm^{-1} but this is no great inconvenience for the investigation of organic compounds. Their performance at wavenumbers above 2000 cm^{-1} is poor and if it is necessary to examine in detail the C—H or O—H stretching region for example, the rock-salt prism must be replaced by one of lithium or calcium fluoride. Similarly the low frequency region below 650 cm^{-1} can only be studied by using prisms of KBr or CsI having limits of 400 and 200 cm^{-1} respectively.

More recently, grating monochromators have come into prominence for routine work. Advances in grating technology have resulted in the production of diffraction gratings of relatively low cost (comparable with that of an equivalent prism). The grating monochromator commonly contains two gratings to provide optimum performance over the range 400—4000 cm^{-1}, although modern inexpensive instruments perform adequately with one.

At the lower end of the range the performance is rather better than that of a rock-salt prism monochromator, and this is maintained right up to the O—H stretching region (3500 cm^{-1}) where the grating is far superior to the prism.

Prism monochromators have to be protected from moisture which causes fogging of the prism. They are normally sealed from the atmosphere with rock-salt windows, desiccated and maintained some 10° above ambient temperature. Such precautions are unnecessary with gratings which have a further advantage in that their performance is less sensitive to changes in temperature.

The detector converts the radiant energy it receives from the monochromator into an electric signal by sensing the heating effect of the infrared radiation. The simplest type of detector is the thermocouple which, when carefully made, is of adequate sensitivity. The Golay detector which senses a pressure change in the gas in a sealed container (caused by the heating effect) is more sensitive than the thermocouple though less robust. The amplified signal from the detector is passed to a chart recorder whose chart is moved in synchronism with the wavelength scan. (The same motor is sometimes used to drive both the chart and the prism or grating.) A continuous record of absorbance versus wavenumber is thus obtained. This is the i.r. spectrum of the sample.

The recorder used in infrared spectrophotometers almost invariably makes use of a 'pre-calibrated' chart paper on which is printed a grid with the absorbance or transmission scale as the vertical axis and the wavenumber scale as the horizontal axis. The chart is fitted either to a drum or to a flat chart carriage. In some instruments the charts are printed as a continuous roll—the so-called flow-chart recorder system. This eliminates possible calibration errors due to careless alignment by the operator.

The precalibrated chart is less accurate than the system in which wavenumber markers are added to the chart by a separate pen actuated by the monochromator scanning mechanism. The loss of accuracy however is relatively slight and is very much out-weighed by the convenience of precalibrated charts. The accuracy can be improved by calibrating each spectrum with bands of some standard material whose wavenumbers have been accurately determined.

Operation of the Spectrophotometer

When both the sample and reference beams are clear, the light intensity falling on the detector is the same whichever beam is entering the monochromator. Although each beam is selected alternately by the sector mirror the output of the detector is effectively constant (a d.c. signal) as illustrated in Fig. 2.2. This d.c. signal is rejected by the amplifier which responds only to a.c. signals.

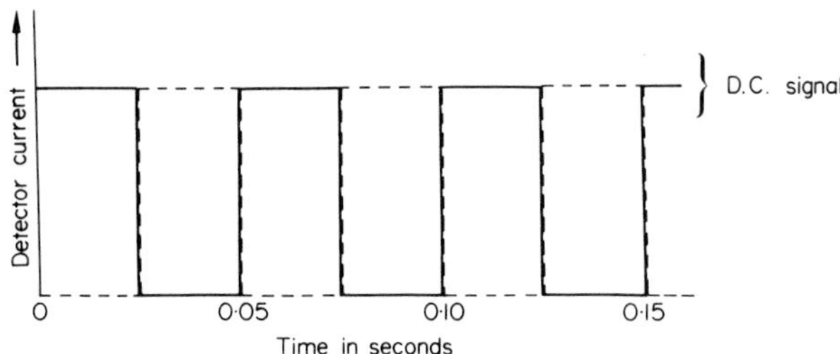

Fig. 2.2. Detector output with no sample in position: —— reference beam falling on detector; - - - sample beam falling on detector.

When the wavelength selected by the monochromator corresponds to that of the radiation which is absorbed by the sample, the signal produced by the sample beam is weaker than that produced by the reference beam. The output from the detector is now effectively an a.c. signal of frequency 20 Hz as illustrated in Fig. 2.3 for the sector mirror chopping twenty times a second.

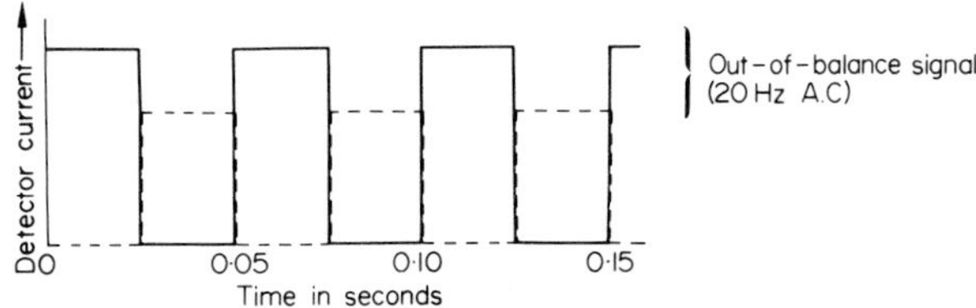

Fig. 2.3. Detector output with absorbing sample in position as sample beam.

The amplifier, which is tuned to receive signals of frequency 20 Hz, accepts the signal and amplifies it. The amplified signal is used to drive an 'optical wedge' into the reference beam, so that the intensity of both beams again becomes equal. The optical wedge is directly connected to the recorder pen and both move to an extent determined by the absorbance of the sample at each wavelength. This arrangement for recording the absorbance is known as the 'optical null' system and is widely used.

The double-beam recording spectrophotometer of the type we have discussed is the only instrument available for infrared spectroscopy. The simple manual spectrophotometer so common in ultraviolet and visible spectroscopy finds no counterpart in infrared spectroscopy where we are usually far more interested in the position of an absorption band than its intensity. In other words infrared spectroscopy is primarily a qualitative technique though it can be used quantitatively when the occasion demands since the absorption of infrared radiation is governed by the same laws that apply in the visible region of the spectrum.

The double beam system has the advantage that variations in source output and detector response with wavelength are automatically compensated and a flat base-line is readily obtained. Atmospheric absorption due to carbon dioxide and water vapour is also compensated and does not appear in the spectrum. When the spectrum of a solution is to be run, the double beam system allows solvent bands to be balanced out by placing an equivalent solvent film in the reference beam. Similarly if the sample has poor transmission characteristics the base line can be restored to near 100% by cutting off some of the reference beam with a screen.

In order to keep the response of the instrument constant over the entire wavelength range it is necessary to ensure that the signal received by the amplifier should be perceptibly constant. This is done by controlling the slit width of the monochromator by means of a 'slit programme' which provides for a wide slit where the source output or detector response is low and a narrow slit where those parameters are most favourable. In some instruments a choice of slit programmes is available, narrow slits giving a higher resolution, and wide slits being used when the transmittance of the sample is low.

Sample Preparation Techniques

When an infrared spectrum of a substance is required, a choice must be made between the various forms in which the material can be introduced into the sample beam of the spectrophotometer.

These comprise:

a solution in a suitable solvent
a thin liquid film (also solid films where suitable, e.g. plastics)
a smooth paste (mull)
a pressed disc with an alkali halide
a gas or vapour
a surface from which the beam is reflected.

In general, one aims to obtain spectral characteristics of 'free' molecules of a compound rather than those for which it is associated with a particular environment (e.g. crystalline lattice). The nearest approximation to the 'free molecule' situation that one can make experimentally is either the gas phase or dilute solution in a non-polar solvent. The need to prepare samples in the gas phase rarely occurs in organic chemistry and the most reliable results are obtained in practice from solution spectra. Despite the advantage however, a solution sample requires time to prepare and, for this reason, it is often more convenient to measure the sample directly, either as a neat liquid or in the solid phase. (Many of the spectra in this book have been obtained this way.) The techniques available for the common sampling methods are given below.

(a) *Solutions.* The concentration normally required for solution spectra is, in general, greater than 0.1 M. In addition to the requirement that a solvent must dissolve the compound at least to that extent, it must be transparent to infrared radiation over the region which is to be examined. Unfortunately, no liquid solvent exists which has no absorption over the entire range covered by modern routine spectrophotometers. In practice, the most satisfactory solvents for infrared spectroscopy are carbon disulphide and carbon tetrachloride, both of which have very few strong bands in the 'rock salt' region. Possible alternatives which dissolve a wider variety of organic compounds are chloroform (or bromoform), methylene chloride and hexane (Fig. 2.5). Thus, in order to obtain the complete spectrum of a compound it is necessary to use two or more solvents separately. In practice a solution of the compound is placed in a fixed path-length cell (Fig. 2.4) which is inserted into the sample beam of the spectrophotometer. A cell of identical path-length, and containing the solvent only, is inserted in the reference beam, thus compensating for the solvent in the sample beam and leading to a spectrum of the solute only. A convenient way to ensure identical path-lengths in both cells involves the use of a variable path-length cell (Fig. 2.6). A fixed path-length cell filled with solvent is placed in the sample beam and the variable path-length cell, also containing solvent, is placed in the reference beam. The spectrometer is set up on a weak solvent band and the variable path-length cell adjusted

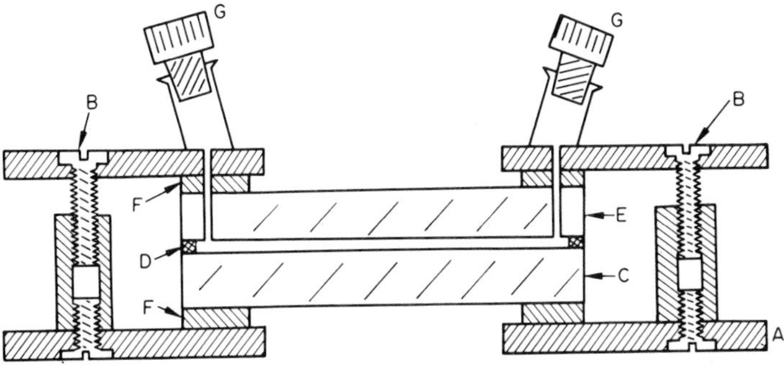

Fig. 2.4. Fixed path-length cell: A—stainless steel frame; B—fixing screws; C—lower rock-salt plate (40 × 22 mm); D—Teflon spacer; E—upper rock-salt plate; drilled for admission of sample; F—neoprene washers; G—filling ports to fit syringe tip; fitted with Teflon plugs.

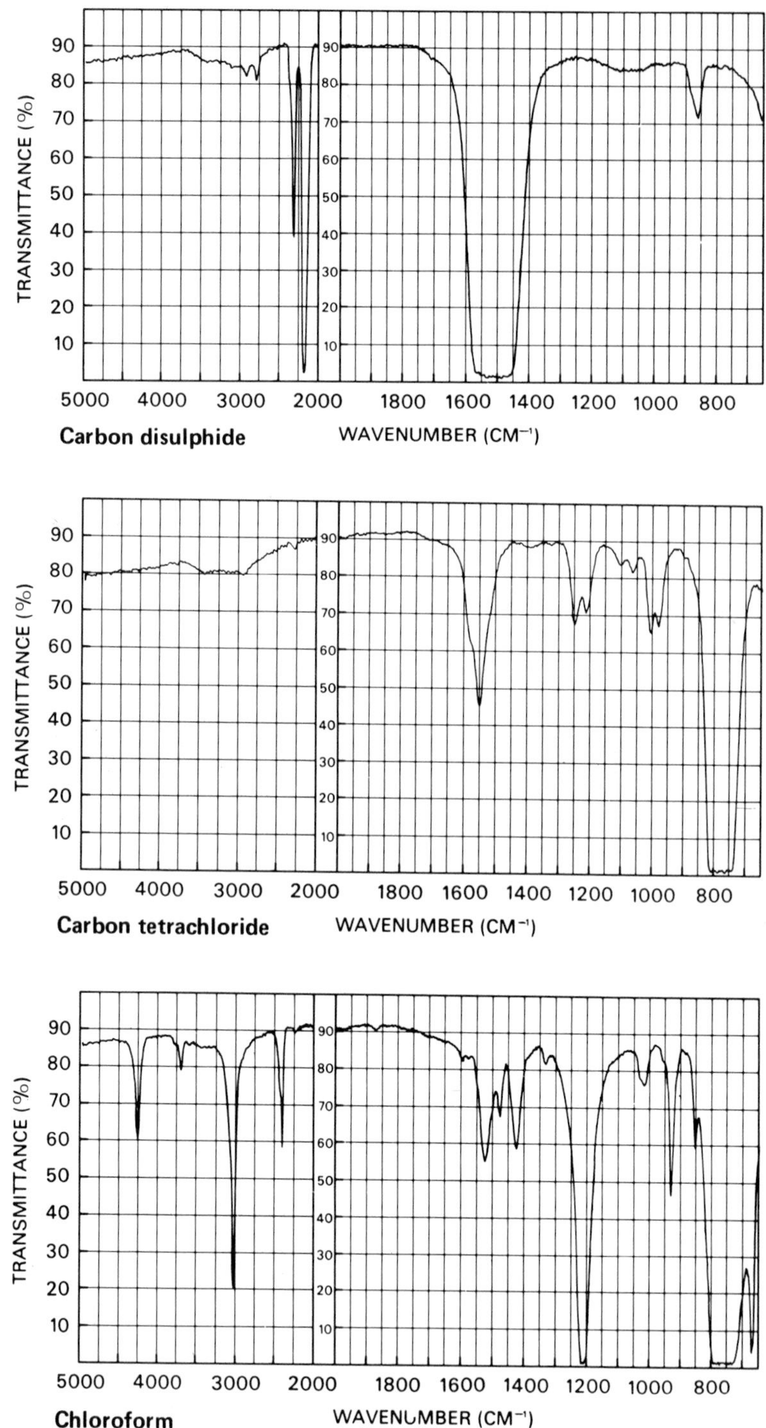

Fig. 2.5. Spectra of some solvents used for infrared spectroscopy (path-length 0.1 mm).

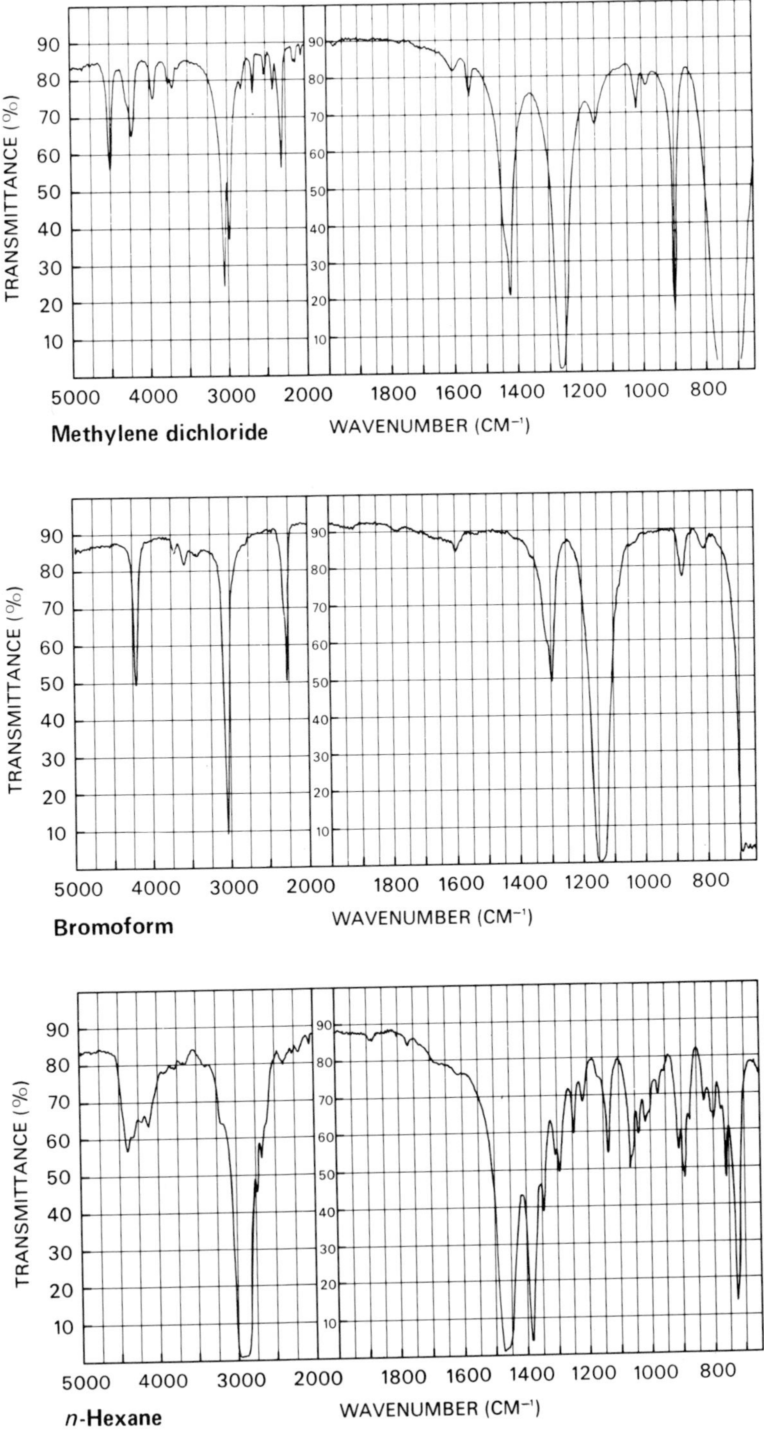

TRANSMITTANCE (%)
90
80
70
60
50
40
30
20
10
5000
4000
3000
2000
1800
1600
1400
1200
1000
800
Methylene dichloride
WAVENUMBER (CM⁻¹)
Bromoform
WAVENUMBER (CM⁻¹)
n-Hexane
WAVENUMBER (CM⁻¹)

Fig. 2.5 *contd.*

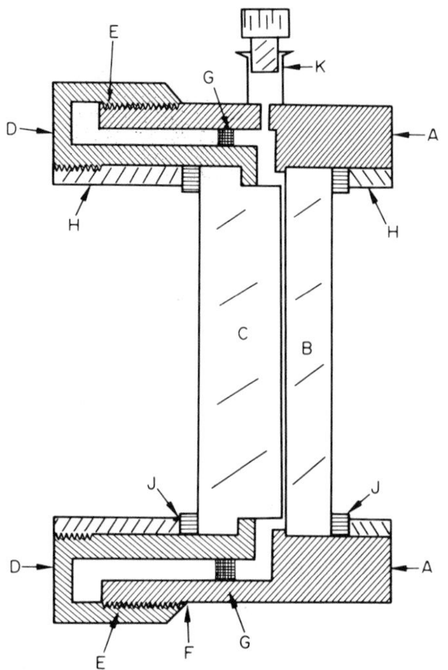

Fig. 2.6. Variable path-length cell: A—cylindrical body; B—back window (32 mm diameter) fixed to body of cell; C—front window (flanged) fixed to rotating inner sleeve; D—Inner sleeve attached to body with micrometer screw; E—micrometer thread; F—micrometer scale; G—Teflon piston seal; H—screwed cylinders retaining windows; J—neoprene washers; K—filling port with Teflon plug.

until no net absorption is observed. The fixed path-length cell is then emptied, filled with the solution and the spectrum measured. Unfortunately, one cannot compensate for strong bands in a solvent as, at these frequencies, all the energy in both beams is absorbed and the instrument becomes totally unresponsive. (It must also be remembered that the 2300 cm^{-1} band of the carbon dioxide in the atmosphere also causes very high absorption in both beams though this is not apparent in the spectrum. Care must therefore be taken when considering bands in this region.)

(b) *Neat liquids*. The quickest way of obtaining a spectrum of a liquid is to place a drop on a rock salt plate and cover it with a second plate to form a thin capillary film or 'contact'. This 'sandwich', mounted in a suitable holder, can be inserted into the sample beam of the instrument. The capillary film is not suitable for volatile materials (b.p. $< 100°C$) because the sample will evaporate in the infrared beam. Under these circumstances the sample must be contained in a fixed path-length cell. Since the liquid film in the cell with the shortest available path-length is several times thicker than the capillary film it will usually be necessary to dilute the sample with a suitable solvent in order to examine the precise frequency and band contour of the strongest bands.

As an alternative to the conventional fixed path-length cell so-called 'Extro-cells' made by extruding silver chloride through a die may be used (Fig. 2.7). These are

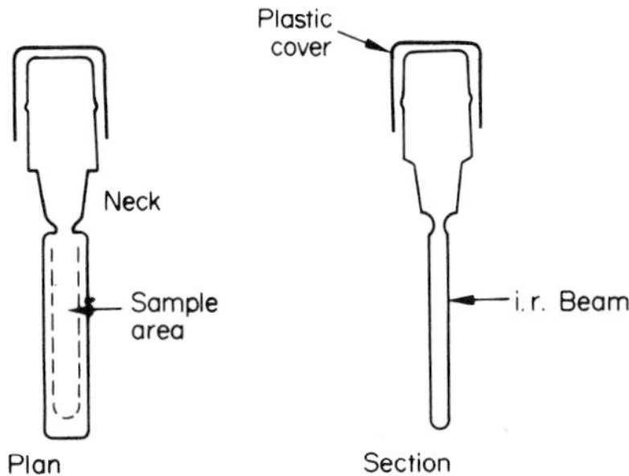

Fig. 2.7. Extrocells.

filled by placing a drop of the sample in the neck and centrifuging it down into the body of the cell, which is then closed with an air tight cover. The Extrocell was first introduced to enable fractions separated by gas chromatography to be collected and their infrared spectra obtained. They are also useful for corrosive samples, very wet samples and samples containing suspended solid particles all of which may damage the rock salt windows of a conventional cell or block its ports. In this application they fulfil the function of a cheap disposable cell with rather greater versatility than other varieties of cell designed for this purpose.

Wherever possible sodium chloride is used as window material for infrared cells, since it is the cheapest and the least easily damaged of the infrared transmitting materials. Its transmission range, 650–4000 cm^{-1}, makes it suitable for the vast majority of applications in organic chemistry. If the spectrum is required at low frequencies potassium bromide (limit 400 cm^{-1}) or caesium iodide (limit 200 cm^{-1}) may be used. High density polythene is also a useful window material for the low frequency region, the remainder of the spectrum being run with rock-salt windows. Another application for which rock-salt is unsuitable is the examination of an aqueous solution. Here silver chloride may be used although calcium fluoride and barium fluoride have better optical qualities. However, the fluorides cannot be used for acid solutions. For such cases, special materials are available (known commercially as 'Irtran').

(c) *Solids.* If the sample does not dissolve in any suitable solvent it will be necessary to examine it in the solid phase. The chief difficulty with solids is the scattering which occurs from the surfaces of the particles, causing the bulk sample to have effectively a very low transmittance. This can be overcome very largely by suspending the solid in a liquid having a similar refractive index. Liquid paraffin (Nujol) is widely used for this purpose, the method being referred to as the 'Nujol Mull technique'. In order to obtain the spectrum the sample is ground up thoroughly in a small agate mortar. One or two drops of Nujol are then added and the grinding continued until a smooth paste is obtained. A small quantity of the paste is transferred to a rock salt plate and pressed out with a second plate to produce a thin film. The 'sandwich' so obtained is mounted in the spectrophotometer in the same way as a liquid contact. Thorough grinding is essential so that the particle size is of the same order as the wavelength of the radiation (a few micrometres).

Inadequate grinding results in poor quality spectra with a high transmission loss at low wavelengths due to scattering. Bands may also be distorted and their apparent centre shifted quite considerably (the Christiansen effect). This adds to the uncertainties associated with the spectra of solids. Hand grinding is usually satisfactory for organic solids but more efficient mechanical grinding in a small mill may be required for the occasional sample which persistently gives poor quality spectra.

Although the spectrum of Nujol is simple, it will obscure the spectrum of the solid in the C—H bending and stretching regions near 1400 and 2900 cm^{-1} respectively. If the C—H vibrations of the solid are of interest they may be observed by using an alternative mulling agent having no C—H bonds for those regions of the spectrum. Hexachlorobutadiene or Fluorolube are possible materials.

The Nujol mull technique is very convenient and is often used as an alternative to the solution technique where high quality spectra are not required, for example in monitoring the progress of a reaction. It must be remembered, however, that the infrared spectrum of a solid may be significantly different from that of the same material in solution or in the liquid state. Some bands may shift by as much as 20 cm^{-1} while splitting may occur in others which are single in the liquid state. Relative intensities will also vary in the different phases. **For this reason one should only compare the spectra of a series of compounds in detail if the spectra were all obtained from samples in solution in the same non-polar solvent.**

Another popular method of obtaining the infrared spectrum of a solid is to grind it up with a suitable alkali halide and subject the mixture to a pressure of some 10 tons/sq. in. in an evacuated die. This produces a clear transparent disc about $\frac{1}{2}$ in. in diameter. Ideally the refractive index of the halide should be as close as possible to that of the sample but, since it would be impracticable to measure the refractive index of all the samples received, for routine work it is usual to employ one halide for all samples. Potassium bromide is most commonly used though some laboratories prefer potassium chloride. The presence of moisture in the disc cannot be avoided when working in the open laboratory but subsequent drying of the disc in a small vacuum oven improves the quality of the spectrum considerably. As with the Nujol mull techniques it is essential that the sample be finely ground to avoid scattering effects. The mechanical mill, for example the type of vibrating mill used to prepare dental amalgam, is to be preferred to hand grinding for the disc technique since, apart from its greater efficiency in reducing particle size, it ensures a uniform mixing of the sample with the potassium bromide. The differences in the spectrum of the solid state compared with the solution are rather more serious with the disc than with the mull owing to the distortion of sample molecules during the preparation of the disc. The presence of ions in the matrix is also a contributing factor and may cause double decomposition to occur if the sample itself is a salt.

(d) *Gases.* The need to obtain spectra from gas phase samples does not often arise in routine structure-determining work. The advent of accessories linking infrared spectrophotometers to gas chromatographs to obtain infrared spectra of samples as they come off the column, however, has increased the number of occasions upon which gas phase spectra are required. The typical gas cell is illustrated in Fig. 2.8.

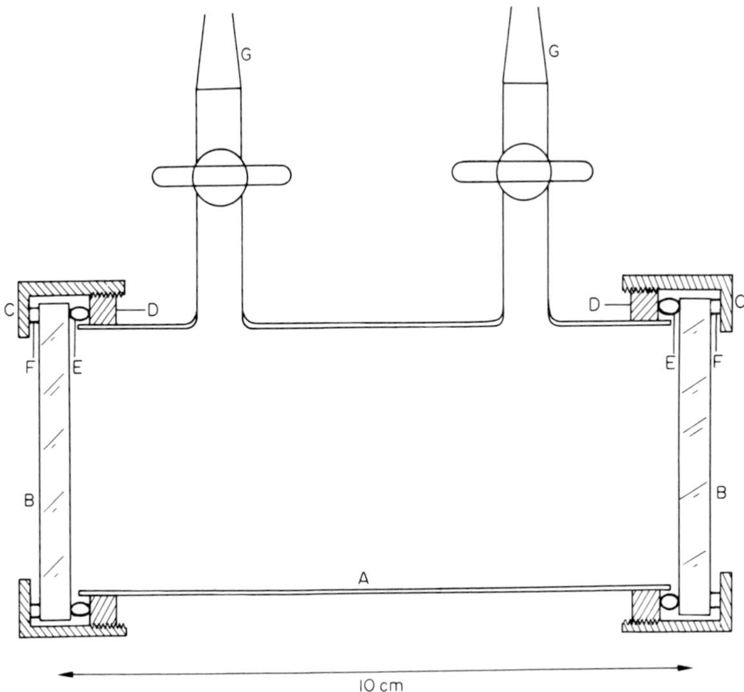

Fig. 2.8. Infrared gas cell: A—Pyrex body; B—rock-salt windows; C—brass end caps; D—threaded brass rings; E—O-rings; F—neoprene rings; G—B.10 cones.

A variant in which the Pyrex body is replaced by a nickel or stainless steel body shaped to fit the beam of the spectrophotometer is known as a **minimum volume gas cell** having a volume some four times less than that of the standard Pyrex cell. For very small samples such as those collected from g.l.c. columns, 'light-pipe' cells having a volume of less than 1 cm^3 are used with a special optical system.

The sample may be introduced into the standard cell by sweeping a current of gas through the cell if large quantities of gas are available. Small quantities can be introduced by vacuum manipulation with the cell attached to a high vacuum system. Very small quantities (<5 cm^3) may be injected with a syringe into an evacuated cell through a septum cap. The spectrum of the vapour of a volatile liquid (b.p. $< 100°C$) may be obtained in the same way or, at atmospheric pressure, by injection through the bore of one of the taps. Vapour spectra can also be obtained from less volatile liquids using a gas cell fitted with a heating jacket.

The spectrum of the gas-phase sample differs from that of the liquid rather more than is the case with the solid. It will, however, bear a much closer relation to the dilute solution spectrum. Bands associated with polar groups such as OH and

CO may shift considerably (100 cm^{-1} or more) due to association in the liquid phase. Hydrogen bonding is particularly noteworthy causing a reduction of at least 200 cm^{-1} in the OH stretching frequency when compared with the gas-phase or solution frequency, together with smaller changes in C—O and other frequencies involving electronegative atoms. The bands of gas samples may also show partially resolved rotational fine structure which will markedly affect their shape and apparent centre when compared with liquid or, in this case, solution spectra.

Reflectance Spectra

Finally, mention must be made of the technique of *Attenuated Total Reflectance* (A.T.R.) which is of increasing importance in many fields. If a beam of i.r. radiation impinges on the back surface of a prism (BC in Fig. 2.9) at an angle of incidence, θ,

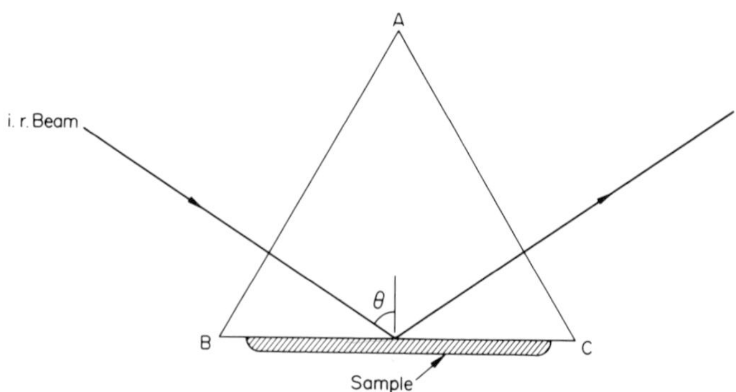

Fig. 2.9. Attenuated total reflectance.

greater than the critical angle, total internal reflection occurs. If this surface is now coated with a sample, the radiation can pass into the sample at those wavelengths where it absorbs while total internal reflection occurs elsewhere. The total internal reflection is therefore attenuated (or 'frustrated') at those wavelengths and the intensity versus wavelength (frequency) curve of the reflected light resembles the absorption spectrum of the sample with some variation in relative intensities and shapes of bands.

The A.T.R. technique is particularly useful for those samples which cannot be examined by the usual sample handling methods and which do not dissolve in a suitable infrared solvent. The following applications are typical.

(a) Viscous liquids, pastes, tars etc, which cannot be pressed out into a thin enough film for transmission purposes.

(b) Aqueous samples—A.T.R. prisms are made of materials such as germanium, silver chloride, KRS 5 (thallium bromide and iodide), which are resistant to water. Furthermore the effective path-length is shorter than is possible by transmission techniques in which the spectrum of water is always very strong.

(c) Plastic films which are too thick for transmission work (>0.1 mm). Such materials are frequently too soft to be ground up and examined by either of the solid techniques. Furthermore the A.T.R technique is 'non-destructive' and records bands which may be correlated with the physical structure of the film.

(d) Surface films and layers including paints. Again the technique permits an *in situ* investigation of the surface layer only (to a depth of $\sim 2\ \mu m$) and may be used to investigate the effects of weathering or chemical action on the surface.

In addition to the prism the 'A.T.R unit' also contains a mirror system to allow the unit to be slipped into the normal sample position of the spectrophotometer. The mirrors direct the beam through the prism and also correct the focussing of the beam to allow for the extra distance through which it passes.

An extension of the A.T.R technique is the Multiple Internal Reflection (M.I.R) system in which a specially shaped prism is used to cause many internal reflections (typically 25) to occur, the intensity of the beam being attenuated at each reflection. Two M.I.R prisms are illustrated in Fig. 2.10. The M.I.R. spectrum is much more intense than that obtained with a single reflection and the multiple units are now almost exclusively used in A.T.R. work.

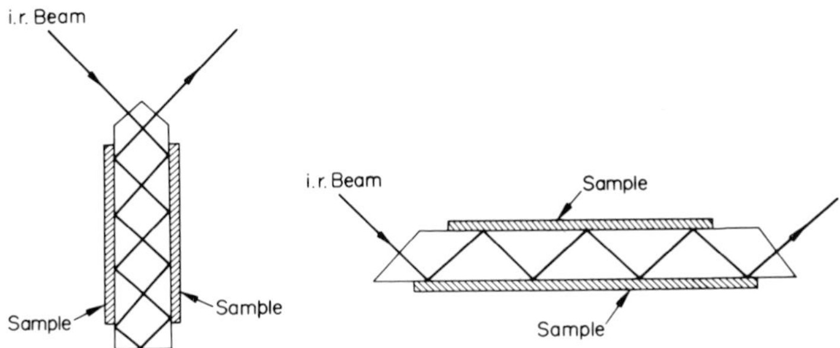

Fig. 2.10. Multiple internal reflection prisms.

Calibration of Spectra

As we noted earlier, the routine infrared spectrophotometer makes use of pre-calibrated chart paper and the accuracy with which frequencies of absorption bands can be measured is subject to calibration errors which may be quite serious. They arise through many causes, chief of which are inaccurate positioning of the chart on the chart carriage or drum, stretching of the chart paper, inaccuracies in the printing of the chart, and deterioration of moving parts in the scanning system or chart drive of the instrument (e.g. wear on cams and pulleys, slackness in bearings, stretching of control wires). It is often of some importance to know the frequency of a band with some accuracy ($\pm 2\ cm^{-1}$) in order to be sure whether or not it falls within a quoted range and also distinguish between the various modifications of a characteristic frequency such as that of the carbonyl group. Such accuracy is also required in the investigation of solvent shifts and effects of substituents on a particular group frequency. A considerable improvement to the accuracy can be made by calibrating the spectrum as soon as it has been run by recording, on the same chart, bands of some reference substance. The frequencies of the bands in the reference must be known accurately and must not vary from one sample to another. Gases (e.g. ammonia) are most satifactory but the spectra tend to be rather too complicated for routine work. Liquids not subject to hydrogen bonding are more satisfactory for routine instruments, indene being a convenient material. Polymer films are most convenient since they can be permanently mounted in a frame which fits the sample position of the spectrophotometer. The film must be reproducible, however, and the polymer used must therefore not exist in different structural forms. Polystyrene has been found quite satisfactory in this respect and provides a large number of reasonably sharp bands. The frequencies of the reference bands are read off the chart and a correction curve of $(\nu_{true} - \nu_{observed})$ versus observed frequency is plotted. This is used to correct the observed frequencies of the sample.

Exercises

1. Under what circumstances would consideration of the experimental technique used be important in the interpretation of an infrared spectrum?
2. Outline the construction and mode of operation of a simple infrared spectrophotometer.
3. Summarize the techniques available for sample preparation prior to taking an infrared spectrum of an organic compound. (You will find this useful for future reference.)

Further Reading

R. T. Conley, *Infrared Spectroscopy*, Allyn and Bacon, Boston, 1966.

A. D. Cross and R. A. Jones, *An Introduction to Practical Infrared Spectroscopy*, Butterworths, London, 1969.

R. G. J. Miller and J. C. Stace (Ed.), *Laboratory Methods in Infrared Spectroscopy*, 2nd Edn., Heyden & Son, London, 1972.

A. E. Martin, *Infrared Instrumentation and Techniques*, Elsevier, Amsterdam, 1966.

B. W. Cook and K. Jones, *A Programmed Introduction to Infrared Spectroscopy*, Heyden and Son, London, 1972.

J. H. van der Maas, *Basic Infrared Spectroscopy*, 2nd Edn., Heyden and Son, London, 1972.

PART 3

CHARACTERISTIC GROUP FREQUENCIES AND THE EMPIRICAL METHOD

Aim

The aim of this part of the programme is to enable the student to recognize the presence of the common functional groups from the spectrum of an organic compound by:

(a) making intelligent use of tables of characteristic frequencies;
(b) applying the fundamental principles covered in the first part and extended in the present one.

New Terms and Concepts

Use of empirical data, skeletal vibrations, fingerprint region.

Prior Knowledge

Part 1, inductive and resonance effects, conjugation, hybridization in carbon compounds, simple organic nomenclature, the structures of simple organic compounds, functional groups, bond order, bond strength and bond length.

Objectives for Part 3

When you have completed Part 3 you should be able to:

1. Use an infrared spectrum and frequency correlation charts to distinguish between a given selection of structural alternatives for a substance or, given other data, to suggest a structure for the substance.

2. Give the approximate stretching frequency (within $200\,cm^{-1}$) of a bond type without reference to the correlation charts.

3. Identify structural features which account for variation in stretching frequency of a given bond (e.g. C=O) from molecule to molecule.

4. Decide from an infrared spectrum of an oxygen-containing compound whether or not its oxygen atom(s) are present as hydroxyl, ether or carbonyl functions (without reference to charts).

5. Distinguish between bands in an infrared spectrum which are reliable for diagnostic purposes and those which are not.

6. Select the most potentially useful absorption bands in the spectrum of an unknown substance for identifying the correct structure from given alternatives.

7. Recognize situations where use of the 'fingerprint region' might be informative and assess correctly the reliability of absorption bands within that region.

8. Identify features in an infrared spectrum of a substance which support or contradict a proposed structure.

PART 3: CHARACTERISTIC GROUP FREQUENCIES AND THE EMPIRICAL METHOD

Introduction

The nature of infrared radiation and the way it interacts with matter to give an infrared spectrum was discussed in Part 1.

A spectrum may, in fact, be regarded as a physical property of the compound which, like its boiling point, melting point and refractive index, may be quoted as a characteristic feature and an index of purity. An infrared spectrum, however, contains much more information than other physical properties, and provides a more reliable means of identification and assessment of molecular structure.

In Part 3, we shall examine in some detail the way in which an infrared spectrum is interpreted, and the structural information extracted. This may be treated as a purely empirical process, and you will soon be able to use aids such as the correlation charts provided, to establish the presence of such groups as OH, C=O, CH, NH_2 without any other information. However, the knowledge of basic theory that you acquired in Part 1 will enable you to understand

(a) why the frequency corresponding to a particular group falls where it does in the correlation chart;
(b) why some characteristic frequencies fall within a narrow range while others cover a much wider range; and
(c) the factors which influence the precise position of the characteristic frequency within the quoted range of a group in a particular molecule.

You will also be prepared for and able to account for any anomalies which may occur.

Multiple Choice Test

As with Part 1, you should attempt a multiple-choice test at this point, making use of the correlation Charts 2 and 4 where appropriate. Write your answers down on a separate sheet of paper and mark the result in the manner described in Part 1.

1. A compound of formula C_5H_8O has a strong infrared absorption band near 1675 cm^{-1} and another near 1600 cm^{-1}. Which of the following isomers is consistent with these data?

(a) $CH_3CH{=}CHCH_2CHO$ (b) $CH_3CH{=}CH{-}\overset{\overset{\displaystyle O}{\|}}{C}{-}CH_3$

(c) cyclic ketone: $CH_2{-}C({=}O)$ bonded to CH_2, ring $CH_2{-}C(=O){-}CH_2{-}CH(CH_3)$, with substituent $CH_3{-}CH$

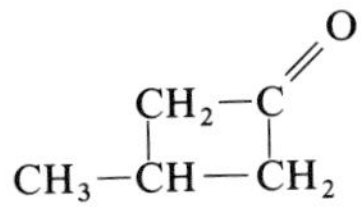

2. Select the most likely frequency for the N=N stretching vibration (**without using the charts**).

(a) 2380 cm^{-1}; (b) 1600 cm^{-1}; (c) 800 cm^{-1}.

3. The compound tropone (I) has a very low carbonyl stretching frequency as a result of a substantial contribution from a polar canonical form. Which of the following would be most appropriate?

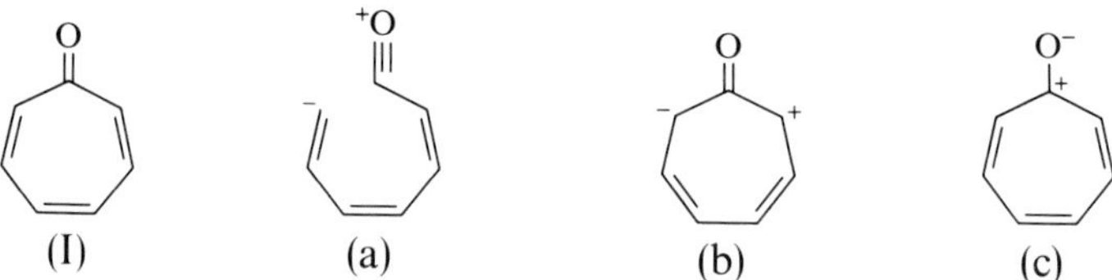

4. A compound was found to have the empirical formula $C_5H_{10}O$ and its infrared spectrum is given in Fig. 3.1. The molecular formula indicates the presence of one double bond **or** one ring. Deduce which of the following assessments about its structure is correct.

   (a) The oxygen atom is present in a carbonyl group and the compound is open-chain.
   (b) The oxygen atom is present as an ether linkage and the compound is cyclic.
   (c) The oxygen is present as an ether linkage but there is a double bond present. This makes it an open-chain system.

5. The absorptions X, Y and Z in the spectrum of *p*-methoxybenzaldehyde, Fig. 3.2, have one of the following features in common. Which one?

   (a) They are all caused by skeletal vibrations.
   (b) They are all caused by stretching vibrations.
   (c) They are all very useful for characterization purposes.

6. Which of the following regions of the spectrum would be most likely to enable you to distinguish between the two isomers below?

   (a) 2500—4000 cm^{-1};
   (b) 1500—2500 cm^{-1};
   (c) 750—1500 cm^{-1}.

$NHCH_3$ CH_3CH_2 $N(CH_3)_2$ CH_3

7. Information regarding the substitution pattern of a benzenoid ring is generally obtained from the study of the molecule's

   (a) C—H stretching vibrations;
   (b) C—H in-plane bending vibrations;
   (c) C—H out-of-plane bending vibrations.

8. Many bands occuring below 1500 cm^{-1} are unreliable as characteristic frequencies because
   (a) extensive coupling of vibrations occurs in this region;
   (b) they are all due to ill-defined bending vibrations;
   (c) they are all highly localized vibrations.

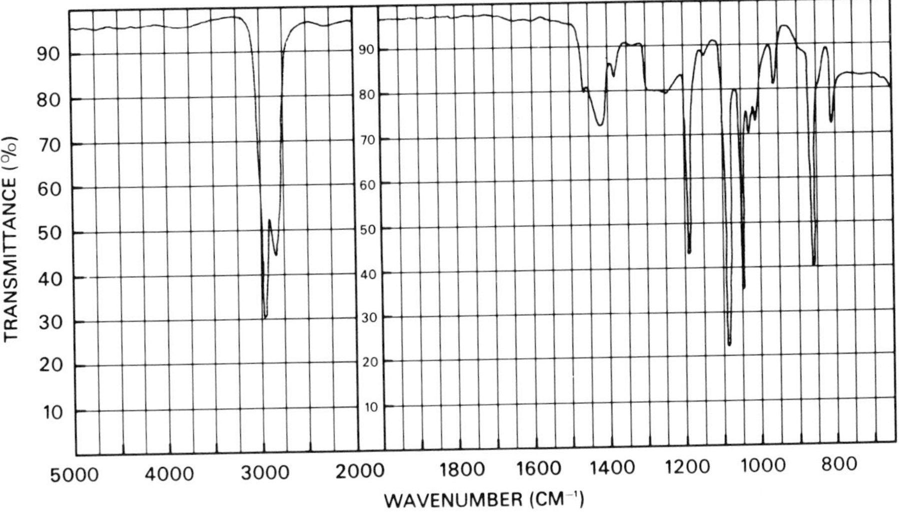

Fig. 3.1. Spectrum of a liquid of formula $C_5H_{10}O$ (Question 4).

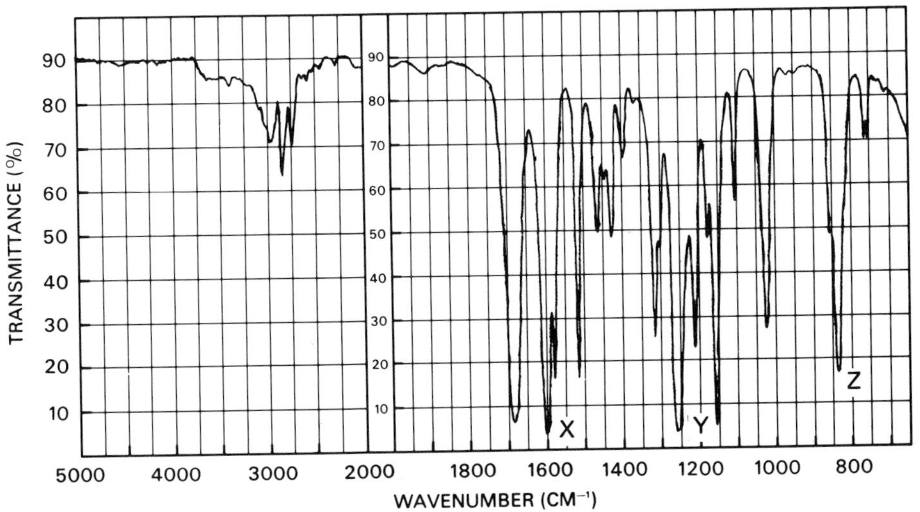

Fig. 3.2. Spectrum of *p*-methoxybenzaldehyde (Question 5).

9. Select the correct order of increasing frequency for the following vibrations **without referring to the charts.**

(a)	C—H	C—H	C=O	O—H	C=C	C≡N
	out-of-plane bend	stretch	stretch	stretch	stretch	stretch
(b)	C—H	C=C	C=O	C≡N	C—H	O—H
	out-of-plane bend	stretch	stretch	stretch	stretch	stretch
(c)	C—H	O—H	C—H	C≡N	C=O	C=C
	out-of-plane bend	stretch	stretch	stretch	stretch	stretch

10. A compound is suspected of having the structure (II). However, its infrared spectrum includes maxima at 3500, 2900, 2700, 1710, 1460, 1150, 830 and 790 cm^{-1}. Two of these absorptions indicate the structure is wrong. They are

(a) 790 and 2900 cm^{-1};
(b) 2700 and 1710 cm^{-1};
(c) 2700 and 1150 cm^{-1}.

(II)

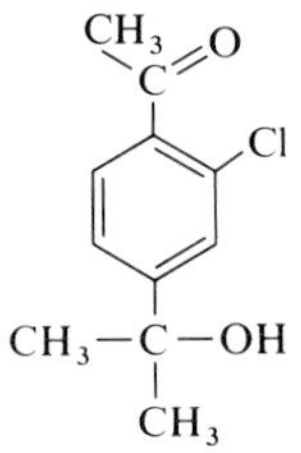

CH_3 O
C
Cl
CH_3—C—OH
CH_3

Comments on marks

0–2 Proceed with Part 3.

3–7 You have some knowledge of the use of infrared spectra, but it would be in your interest to secure a more thorough understanding and you are advised to work through Part 3—it shouldn't take you long.

8–10 Unless you obtained full marks, consult the sections of Part 3 relevant to your incorrect answers; then proceed directly to Part 4.

The Empirical Use of Infrared Frequencies

F82 In Part 1 we saw that the infrared spectrum of a compound may be interpreted in terms of the vibrational motions of the molecules. The mechanical frequency of a vibration is determined by the strengths of bonds and the masses of moving atoms. By a process akin to resonance in acoustics, the presence of such a vibration results in the absorption of infrared radiation of the same frequency by the molecules.

Some vibrations are located almost entirely within a small group of atoms, the remainder of the molecule being unaffected except to a very small extent. We shall, therefore, expect all molecules containing the same group of atoms to absorb infrared radiation at *precisely the same frequency/approximately the same frequency/quite different frequencies* provided there is a vibration which essentially involves only that particular group.

F83 This is the basis of the use of infrared spectroscopy for identifying FUNCTIONAL GROUPS by the presence of features such as the characteristic frequencies of those groups in the infrared spectrum of a compound.

The functional groups present in compound (**1**) can be listed as

* a saturated hydrocarbon fragment (aliphatic C—H) CH_2—CH, CH, CH_2, CH_2

a benzenoid ring,

a hydroxyl group, —OH

List the functional groups present in the other three compounds.

OH, CH_2 — CH, CH, CH_2, CH_2 (attached to benzene ring)

(**1**)

CH_3CH_2OH

(**2**)

CH_2CHO (attached to benzene ring)

(**3**)

OH, CH, C, CH_2, CH_2, $CH=CH_2$, CH_2, CH_2, CH_2, N

(**4**)

* For the purpose of infrared spectroscopy hydrocarbon fragments are regarded as 'functional groups'.

A82 approximately the same frequency.

F84 Characteristic absorption frequencies associated with all the functional groups commonly found in organic molecules are now well-known as a result of systematic studies of the infrared spectra of series of compounds, each series containing a common structural element.

In order for a band in the infrared spectrum to be characteristic of a functional group it must be

(a) present in the spectra of all molecules containing the group;
(b) absent from the spectra of all molecules not containing the group.

This empirical test is in fact the only way of assessing the reliability of any characteristic frequency. It does not depend upon the theory of infrared spectroscopy, but a knowledge of the theory does enable the results obtained to be rationalized, and leads to an understanding of trends and minor variations.

In the following table the frequencies of the main features in the spectra of eight compounds are shown. Deduce the frequency range characteristic of the hydroxyl group.

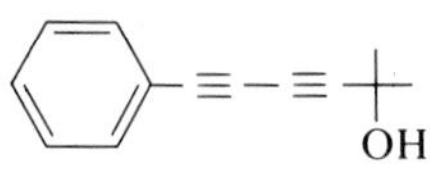

OH

(5)

1170, 760, 3600, 740, 2950, 950

(6)

2990, 950, 1780, 1160

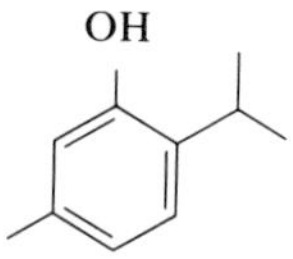

OH

(7)

1150, 2950, 810, 1300, 3600

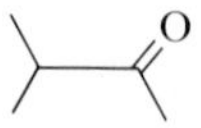

O

(8)

1350, 2980, 1720

N OH

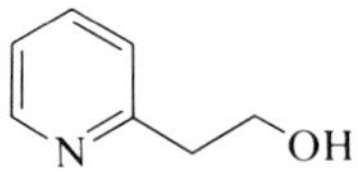

(9)

1440, 750, 2970, 3500, 1125

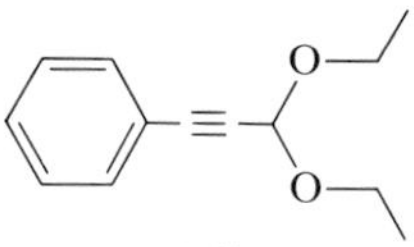

O O

(10)

1120, 750, 2980, 2920, 1450, 1050, 690

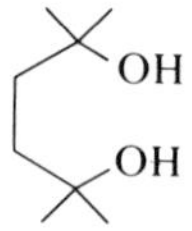

OH OH

(11)

2980, 3600, 1420, 900, 1440

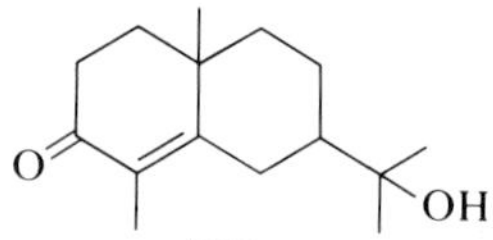

O OH

(12)

2920, 1610, 1020, 1200, 3600, 1670, 1350

A83 In (**2**): a saturated hydrocarbon fragment CH_3CH_2-
a hydroxyl group $-OH$

In (**3**): [benzene ring with methyl], $-CH_2-$, CHO

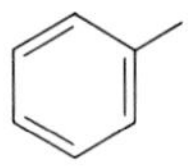

In (**4**): saturated hydrocarbon fragment,

C C C N (C—N with three carbons), $-OH$, $-CH{=}CH_2$

F85 Thus we can detect immediately the presence of features such as the hydroxyl group, carbonyl group, or aromatic ring, in an unknown compound. Moreover we can extend empirical correlations to tell, for example, whether a hydroxyl group is that of an alcohol or a carboxylic acid or whether a carbonyl group is that of a ketone or an amide and, in many cases, whether an aromatic ring is mono-, di- or tri-substituted.

As with many other physical methods it is found in practice that infrared spectroscopy is of widest application as an analytical tool when it is used empirically. The empirical information is usually presented in the form of a chart or table, a typical example of which is given in Chart 2 on the sheet in the pocket at the back of the book. Refer to Chart 2 and assign the following statements as true or false:

(a) C=C stretching frequencies are generally higher than C=O.
(b) The C—D stretching frequency is lower than C—H.
(c) The moderate to intense absorptions of benzenoid compounds which occur between 650 and 900 cm^{-1} are due to the out-of-plane bending of the aromatic C—H bond.
(d) Hydrogen bonding lowers the stretching frequency of an O—H bond.

(e) Dimethylacetamide (**13**) has no fundamental absorption frequencies above 3000 cm^{-1}.

$$CH_3-\overset{\overset{\displaystyle O}{\|}}{C}-N\begin{matrix}\diagup CH_3\\ \diagdown CH_3\end{matrix}$$

(**13**)

A84 3500—3600 cm^{-1}.

F86 Which of the following functional groups is present in the compound whose infrared spectrum is illustrated in Fig. 3.3?

(a) $CONH_2$ (b) aliphatic C—H (c) C≡N (d) S—H
(e) C—D (f) aromatic ring (g) NH_2

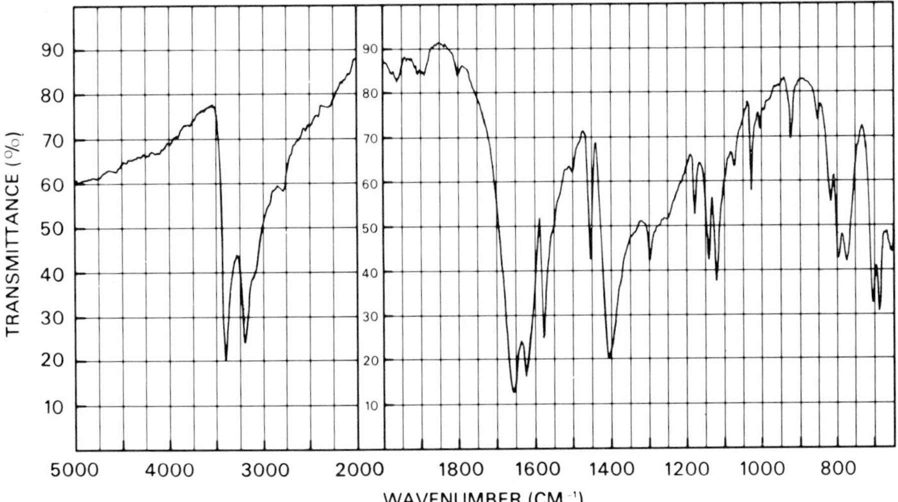

Fig. 3.3. Spectrum of a solid run as a KBr disc.

A85 (a) false (b) true (c) true (d) true (e) true

F87 Which of the following structures corresponds to the infrared spectrum shown in Fig. 3.4?

(**14**) cyclopentanone (ring with C=O)

(**15**) NH_2 on benzene ring

(**16**) $H_2C=CH-CH=CH_2$

(**14**) (**15**) (**16**)

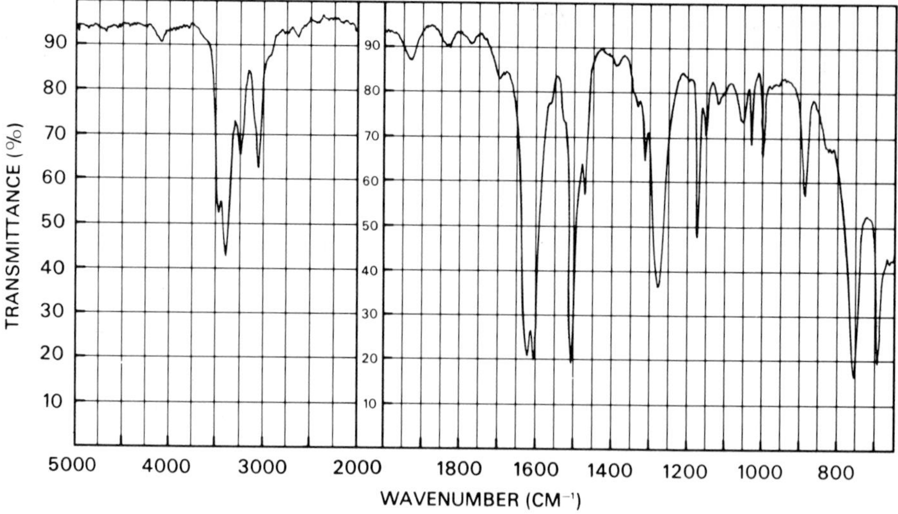

Fig. 3.4. Spectrum of a liquid run as a capillary film.

A86 (a), (f), (g).

F88 Which of the following structures corresponds to the infrared spectrum shown in Fig. 3.5?

$CH_3COOC_2H_5$	CH_3CONH_2	CH_3COCl
(17)	**(18)**	**(19)**

A87 (**15**). Note the absorption due to NH stretching above 3000 cm^{-1}, the prominence of bands due to a benzenoid ring, and no saturated C—H.

F89 Which of the following structures corresponds to the infrared spectrum shown in Fig. 3.6?

CH_3 CH_3 CH_3 CH_3 CH_3 CH_3 CH_3 CH_3 CH_3

(20) **(21)** **(22)**

A88 (**17**). There is no NH_2 absorption. Strong absorption between 1000 and 1300 cm^{-1} indicates the presence of C—O.

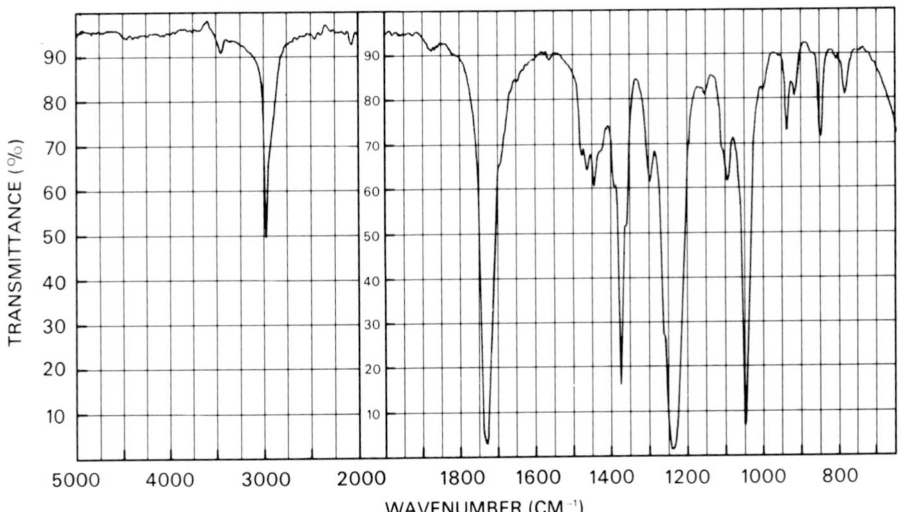

Fig. 3.5. Spectrum of compound in solution in CS_2 (650—1000 cm^{-1}) and CCl_4 (1000—5000 cm^{-1}) (**F88**).

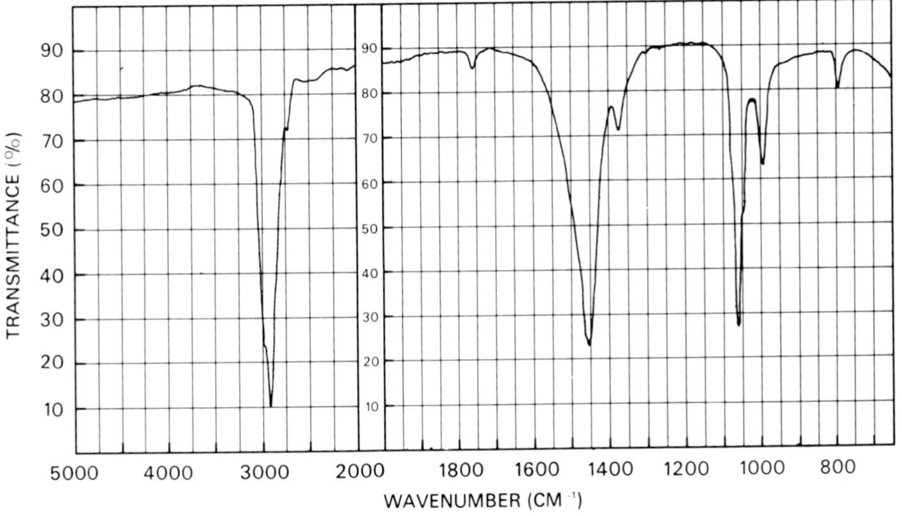

Fig. 3.6. Spectrum of solid run as KBr disc (**F89**).

A89 (**22**). (**20**) and (**21**) would show absorption due to out-of-plane aromatic C—H bending. The feature near 800 cm^{-1} is much too weak for this. Also, notice how the symmetry of (**22**) leads to a very simple spectrum.

Trends in Frequencies

(i) Single bond stretching frequencies

F90 Notice that you were able to correlate a spectrum with a structure in each of the last four frames purely by using the information in the correlation chart.

This is an empirical process requiring no knowledge of the theory of infrared spectroscopy but uses information collated by the method described in **F84**. Although many structural problems can be solved in this way, a more penetrating use of infrared spectroscopy is available to those who understand the reasons for the relative position of groups in the chart. The provision of the ground work for such an understanding was one of the main aims of Part 1.

Let us now consider selected areas of Chart 2 and see if we can rationalize the trends observed. First, notice how the frequencies of the common stretching vibrations are distributed. One set of these vibrations have noticeably higher frequencies than all the rest. Which of the following groups are included in this set?

C=O C—C C—H N=O S—H N—H C—O O—H

F91 It is helpful now to return to our model of two balls joined by a stretched spring. Which of the following statements gives the reason why OH, CH, NH, SH stretching vibrations have so much higher frequencies than other groups?

(a) Hydrogen is a very light atom and the other atom is relatively heavy.
(b) Hydrogen bonding is involved.
(c) There are considerable electronegativity differences between the atoms involved.

- - - - - - - - - - - - - - - - - - - -

A90 C—H, S—H, N—H, O—H.

F92 In Part 1, you calculated the vibration frequency of HCl and found it to be 2886 cm^{-1}. The aliphatic C—H bond vibrates at almost the same frequency in spite of the fact that the mass of chlorine is three times the mass of carbon. Bromine has twice the mass of chlorine. With this in mind select the most likely value for the stretching frequency of the hydrogen bromine bond from those given below:

2863, 2935, 2559 cm^{-1}.

- - - - - - - - - - - - - - - - - - - -

A91 (a). Hydrogen bonding has only a secondary effect on frequency. Many lower frequency vibrations are characteristic of bonds between atoms of very different electronegativities.

F93 Although you would expect the value to be very close to that for HCl, (and indeed, 2863 cm^{-1} is the calculated value based on mass differences alone.) the observed value is, in fact, 2559 cm^{-1}. The apparent discrepancy arises because mass is only one of two factors which determine frequency. The calculation referred to above neglected the other. What is the other?

- - - - - - - - - - - - - - - - - - - -

A92 2863 cm^{-1}. Once again the very small mass of the hydrogen atom is the dominating feature.

F94 Deuterium (D) has twice the mass of the normal isotope of hydrogen, protium (H). By using the usual relationship

$$\omega = \frac{1}{2\pi}\sqrt{\frac{k}{m_1 m_2/(m_1 + m_2)}}$$

select the most appropriate value for the DCl stretching frequency, ω_{DCl}, from the values

2860, 2100, 1400 cm^{-1}.

Note that in the present case of comparison, where we can assume identical force constants, the expression reduces to

$$\omega_{DCl} = \omega_{HCl}\sqrt{\frac{m_H}{m_D} \times \frac{m_{Cl} + m_D}{m_{Cl} + m_H}} \qquad \begin{cases} m_H = 1 \\ m_D = 2 \\ m_{Cl} = 35 \end{cases} \qquad \sqrt{2} = 1.4$$

A93 The force constant (k) which is closely related to the strength of the bond.

F95 You will notice that doubling the mass of the light atom has a much greater effect on the frequency than much larger changes in the mass of the heavy atom. The strength of the carbon–deuterium (C—D) bond is virtually the same as that of the carbon–protium bond (C—H). Without looking at the chart, state whether you think the C—D stretching absorption will be observed at higher or lower frequency than the C—H stretching absorption.

A94 2100 cm^{-1}.

$$\omega_{DCl} = 2886\sqrt{\left(\frac{1}{2} \times \frac{35 + 2}{35 + 1}\right)} \approx \frac{1}{\sqrt{2}} \times 2886\ \mathrm{cm}^{-1}.$$

F96 Look at the chart and write down the frequency range of the C—D stretching vibration.

A95 Lower. Compare ω_{DCl} and ω_{HCl}. As with X—H vibrations, all X—D vibrations have similar frequencies.

F97 Consider the C—H stretching vibration. Which will cause the greater changes in frequency, exchanging carbon for chlorine or hydrogen for deuterium?

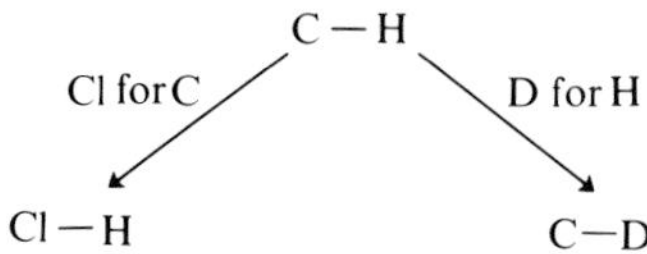

A96 2100—2300 cm^{-1}.

F98 The deuterium atom is still very small compared with carbon so it is not surprising to find that C—C stretching vibrations occur at much lower frequencies. Write down the range over which they may be found. (Refer to Chart 2).

A97 hydrogen for deuterium. See **F95**.

F99 Although the oxygen and nitrogen atoms are a little heavier than the carbon atom, the difference is small compared with the difference between carbon and hydrogen. Therefore, it is not surprising to find that C—O and C—N absorb in roughly the same broad region as the C—C stretching bands. Look at the chart and write down the range over which the stretching frequencies of the C—C, C—O and C—N bonds overlap.

A98 750—1200 cm^{-1}.

A99 1020—1200 cm^{-1}.

(ii) Multiple bond stretching vibrations

F100 Refer again to Chart 2 and write down all the functional groups which have characteristic stretching frequencies in the region 1500—2400 cm^{-1}. With the exception of C—D, what structural feature do all these groups have in common?

F101 Now find the ranges for the corresponding groups in which the same atoms are connected by a single bond. Arrange the following groups in order of increasing frequency:

(a) C=C C—C C≡C;

(b) C≡N C=N C—N.

A100 All the groups contain a multiple bond.

F102 In both cases we see that the frequency *increases/decreases* with increasing bond order.

- - - - - - - - - - - - - - - - - - - -

A101 (a) C—C < C=C < C≡C;

(b) C—N < C=N < C≡N.

F103 Thus stretching frequencies of bonds between atoms of similar mass are determined almost entirely by the bond order, with the result that a given range of the spectrum is characteristic of triple, double or single bonds (excluding bonds to hydrogen) and independent of the atoms linked.

Carbon monoxide absorbs radiation at a wavenumber of 2143 cm^{-1}. What does this suggest to us about the bond in carbon monoxide?

- - - - - - - - - - - - - - - - - - - -

A102 increases. Increasing bond order parallels increasing bond strength and decreasing bond length.

F104 In fact the majority of multiple bonds absorb above 1500 cm^{-1} in the infrared, and are generally sufficiently separated from other absorptions to be unambiguously identified. Indeed, the general point can be made that absorptions above 1500 cm^{-1} are the most diagnostic and useful bands in an infrared spectrum. Which of the following groups of vibrations includes **only** those which absorb above 1500 cm^{-1}?

(a) C—C, C—O, C—N, and multiple bond stretching vibrations.
(b) C—C, C—O, C—N and O—H, C—H, N—H stretching vibrations.
(c) Multiple bond and O—H, C—H, N—H stretching vibrations.

- - - - - - - - - - - - - - - - - - - -

A103 It is a triple bond.

A104 (c).

(iii) Skeletal vibrations

F105 Now let us consider the region of the spectrum below 1500 cm^{-1}. Refer to the chart and write down three categories into which we may classify all the types of vibration found in this region.

F106 The double bond stretching vibrations are diagnostic. Consider the other two categories, starting with the bending vibrations. These invariably absorb at lower frequency than corresponding stretching vibrations. In fact, practically all bending vibrations occur below 1600 cm^{-1}. In the first section of this programme we remarked on their general complexity, there being several types for each group of atoms. As a result, relatively few bending vibrations are sufficiently diagnostic for identification purposes. Indeed the only useful characteristic bending vibrations are those involving hydrogen atoms, In order to remind yourself of these, sketch the different modes of vibration for the $-CH_2-$ group which give rise to absorption below 1700 cm^{-1}.

A105 Double bond stretching **not** involving carbon, single bond stretching and bending vibrations.

F107 Look at the chart and write down the approximate regions of CH stretching frequency and the rocking/wagging and deformation bending frequencies of the CH_2 group.

A106

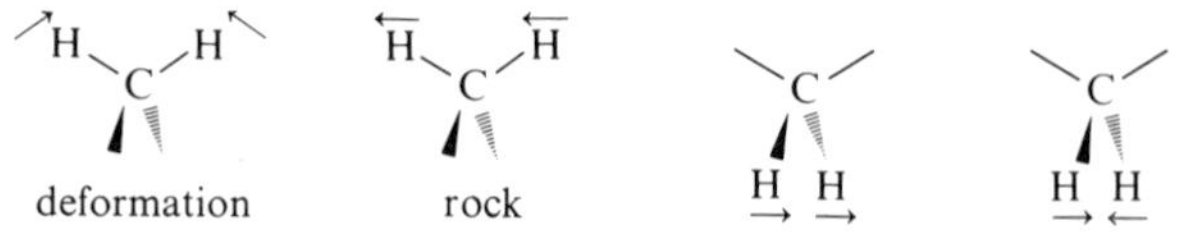

deformation — rock — wag — twist

F108 Bending vibrations which do not involve hydrogen atoms, such as (I) and (II) occur beyond the lower frequency limit of simple infrared instruments and furthermore, they are ill-defined for practically all organic compounds. These vibrations, together with similarly ill-defined C—C, C—O and C—N stretching modes, (**F99**) are commonly referred to as SKELETAL VIBRATIONS since they involve distortion of the molecular framework. Which of the following statements are true?

(I) — (II) — (III) — (IV)

(a) Vibration (III) has a lower frequency than (IV).
(b) Both vibrations (III) and (IV) are ill-defined in the spectra of the majority of organic compounds.
(c) (III) is a skeletal vibration but (IV) is not.

A107 stretching, 2830—3010 cm^{-1}; wagging/rocking 690—1210 cm^{-1}; deformation, 1330—1480 cm^{-1}.

F109 The following vibrations occur in ethylenediamine:

$$NH_2-CH_2-CH_2-NH_2$$

(a) NH stretch (d) CH stretch (f) C—C stretch
(b) CN stretch (e) NH_2 bend (g) CH_2 bend
(c) CCN bend

Select those which can be termed skeletal vibrations.

A108 (a), (c).

F110 Vibrations lying above 1500 cm^{-1} give the most reliable characteristic frequencies since they are localized almost entirely within a single small group of atoms. The skeletal vibrations, on the other hand, are not so localized and extend over the entire framework of the molecule. Thus, if we have a chain of four atoms in a molecule such as ethylenediamine N—C—C—N there will be three stretching vibrations, but each involves changes in all three bond lengths and it is incorrect to refer to them (as we did in **F109**) as C—N and C—C stretching.

Using the chart and the knowledge you have gained in the last few frames, assign appropriate vibrations to the labelled bands in the spectrum of ethylenediamine given in Fig. 3.7.

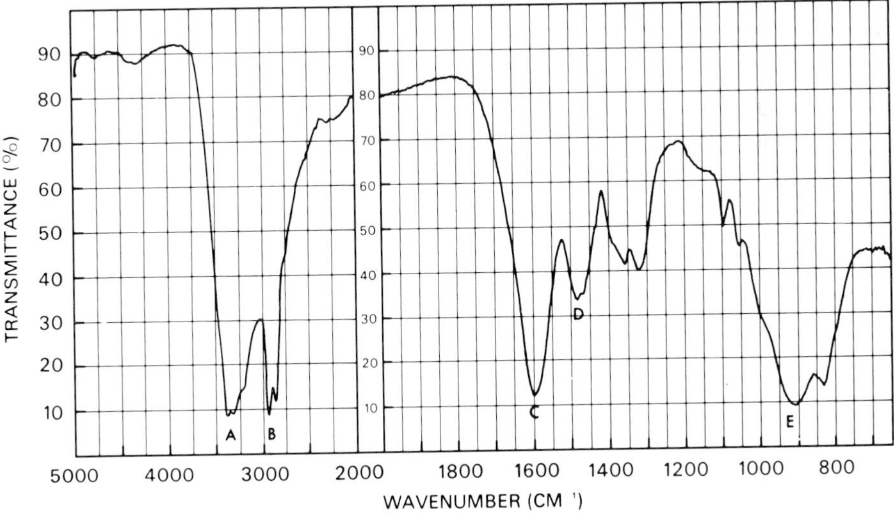

Fig. 3.7. Spectrum of ethylenediamine run as a capillary film.

A109 (b), (c) and (f).

F111 The coupling of the idealized stretching vibrations involving only the C—C or each of the two C—N bonds in ethylenediamine to give three vibrations each involving all three bonds, is typical of skeletal vibrations in general. Such behaviour is very common and may occur whenever vibrations which involve adjacent bonds or groups of atoms have closely similar frequencies.

Refer to Chart 2 and write down the groups which you would expect to couple with a C—O stretching vibration when adjacent to this bond in a molecule.

A110 A. NH stretching B. CH stretching
C. NH bending D. CH bending
E. C—C/C—N skeletal together with CH rocks, wags and twists

F112 The extent to which coupling between adjacent vibrations of similar frequency may occur depends also on the direction of the motion in the two groups. If the motion is in the same direction in both groups, coupling will be considerable. If the motion of one group is perpendicular to the other, then coupling will not occur at all. There are, then, three main factors which determine whether coupling will occur:

(i) the groups involved must be adjacent;
(ii) the natural frequencies of the vibrations must be similar;
(iii) the vibrations concerned must involve motions in the same direction.

With these considerations in mind, select the vibration of *n*-butane (**23**) from the following list which you would expect to couple most strongly with C—C stretching modes.

(**23**)

(a) C—H stretching
(b) HCH deformation of of $>CH_2$
(c) CH_2 rock
(d) CH_2 wag
(e) CCC deformation
(f) CH_3 twisting

A111 Your list should have included C—C, C—F and C—N stretches, CH rocks and wags, but no others.

F113 Thus we find that all the vibrations involving groups having natural frequencies in the range 900—1400 cm^{-1} will, in any given molecule, be coupled to

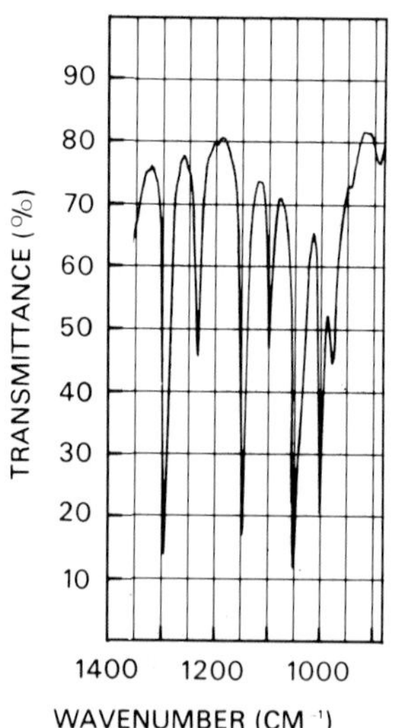

Fig. 3.8. Fingerprint region of spectrum of an unknown liquid.

an extent which depends critically on the shape of the molecule. This coupling accounts for the wide range which we observe when we look for vibrations charac-

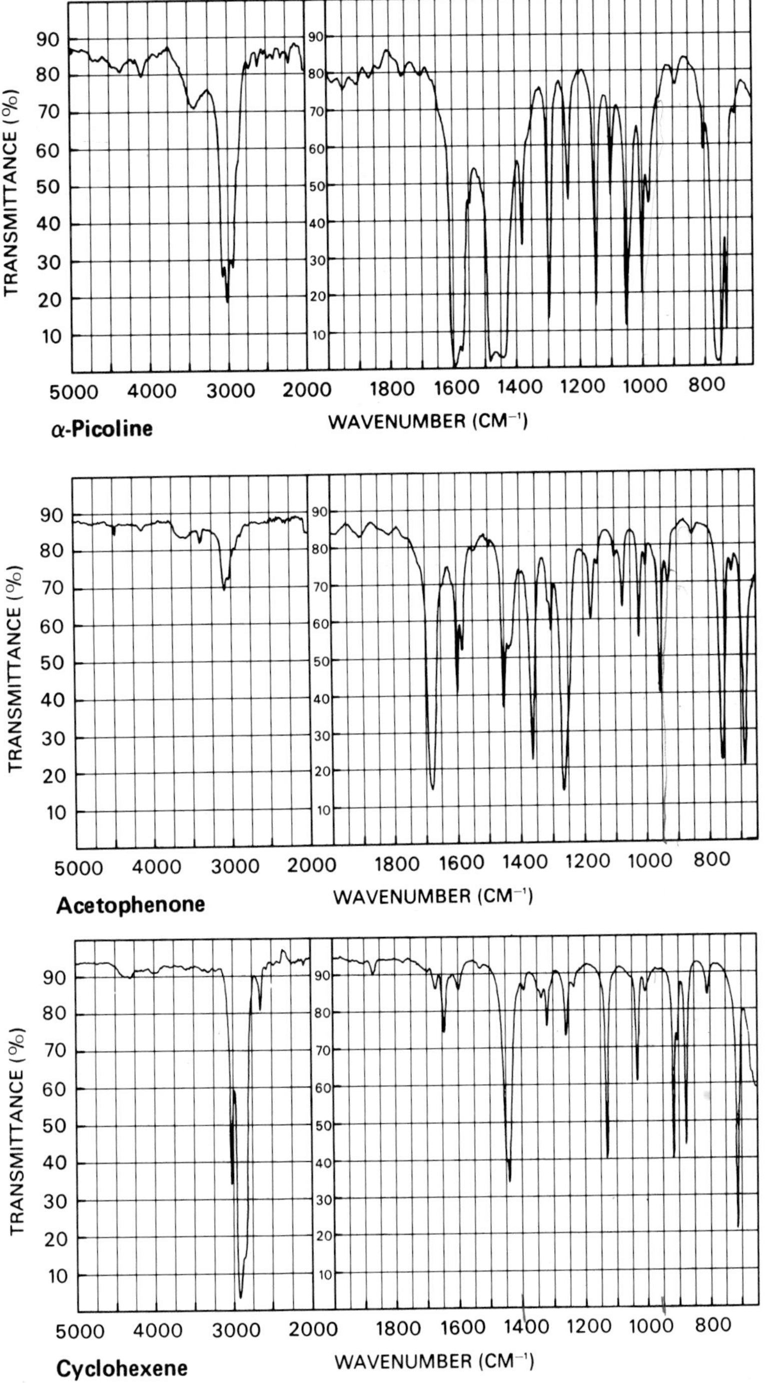

Fig. 3.9. Reference spectra of liquids run as capillary films.

teristic of the C—O bond, for instance. Conversely, one can make the generalization that reliable characteristic frequencies are those which do not couple extensively with other vibrations in the molecule.

The most important practical consequence of the dependence of coupling on molecular shape, however, is that the pattern of the bands in the 900—1400 cm^{-1} region of the i.r. spectrum will differ markedly from one molecule to another, even for molecules having the same functional groups. Indeed, the complexity of the spectrum between 900 and 1400 cm^{-1} makes it unlikely that any two compounds will show exactly the same pattern. For this reason the region is frequently used as a 'fingerprint' for the compound and the range of frequencies involved known as the **fingerprint region.***

Which of the compounds whose complete i.r. spectra are given in Fig. 3.9 has a fingerprint region matching that shown in Fig. 3.8?

A112 (d). CH_2 wag. CH stretch, C—C—C bend and CH_3 twisting do not comply with requirement (ii) and other CH bending vibrations involve movement at right angles to C—C stretching (point (iii)). All vibrations comply with requirement (i).

F114 Which of the following statements is false?

(a) Skeletal vibrations involve both bending and stretching vibrations.
(b) The extent to which two vibrations couple is independent of their relative frequency.
(c) The fingerprint region of the spectrum of *n*-butane will contain some absorptions due to skeletal vibrations.

A113 α-picoline.

A114 (b). See **F112.**

Characteristic Frequencies below 1500 cm^{-1}

F115 Which is likely to have the strongest absorption band?

(a) A skeletal stretching mode involving a C—C bond.
(b) An analogous mode involving a C—O bond.

F116 Although absorptions associated with C—O stretching occur in the fingerprint region and have rather variable frequency, they are usually very much stronger than analogous C—C bands. Indeed the absence of a strong absorption

* Unfortunately the precise range of frequencies included in this term varies according to author.

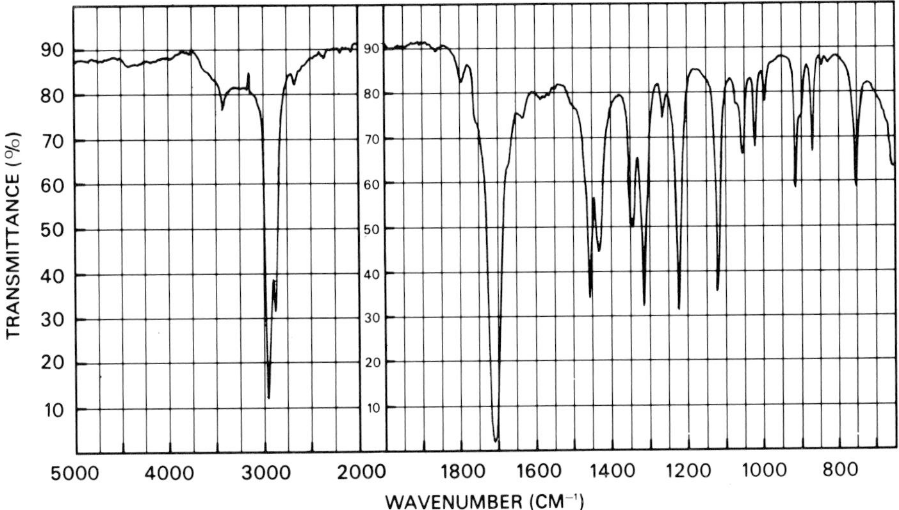

Fig. 3.10. Spectrum of liquid run as a capillary film.

in the fingerprint region usually precludes the presence of a C—O bond in the compound being examined. Which of the four compounds below is responsible for the infrared spectrum in Fig. 3.10?

O= O   O=   O C $CH_3 \cdot O$ O

(24) **(25)** **(26)** **(27)**

A115 (b). C—O stretching involves a larger change in dipole moment.

F117 Similarly C—Cl can usually be identified by the intensity of its stretching band. Unlike C—O, C—C and C—N, however, its frequency range is sufficiently narrow for identification purposes also. Use the chart and the contents of the last three frames to pick out the correct statement from those given below.

(a) 1,2-dichloroethane will have a strong absorption associated with C—Cl stretching near 1000 cm^{-1}.
(b) 1,2-dichloroethane could absorb strongly anywhere between 600 and 800 cm^{-1}.
(c) 1,2-dichloroethane will have a weak absorption associated with C—Cl stretching somewhere between 600 and 1100 cm^{-1}.
(d) 1,2-dichloroethane will have a weak absorption associated with C—Cl stretching near 800 cm^{-1}.

A116 (**25**). There is no band in the 1000—1400 cm^{-1} region strong enough to arise from a C—O group. All compounds except cyclohexanone (**25**) contain a C—O bond. (C—O bands are generally stronger than the CH deformations near 1400 cm^{-1}. In addition they are often quite broad.)

F118 We saw in Part 1 that although in general, bending vibrations are complicated and difficult to describe, the out-of-plane vinyl and aryl C—H bends are well defined. It is not surprising, therefore, that these vibrations give rise to absorptions in the infrared which are useful in recognizing the presence of unsaturation. Which of the following molecules will show such absorptions?

$CHCl_3$ **(28)** $CH_2{=}CCl_2$ **(29)** C_6H_5—CH_3 **(20)**

$C_6(CH_3)_6$ **(30)** $(CH_2)_6$ **(31)** pyrrole (H–N ring) **(32)**

- -

A117 (b). In (a) 1000 cm^{-1} is outside the range; in (c) and (d) C—Cl bands are usually strong.

F119 Bands due to out-of-plane hydrogen bends occur in the region 650—1000 cm^{-1} and are frequently strong and sharp. Indeed, in the case of aromatic compounds they are often the most intense bands in the spectrum. Their frequency range falls below that of most skeletal stretching vibrations and this fact, together with their high intensity, enables them to be identified with reasonable certainty.

Write down the frequencies of bands in Fig. 3.8 which could be attributed to such CH bending modes in the spectra of α-picoline, acetophenone and cyclohexene.

- -

A118 **(29)**, **(20)** and **(32)**. In **(28)**, **(30)** and **(31)** there are no hydrogen atoms attached directly to a double bond.

F120 Furthermore, since the direction of motion in out-of-plane bending vibrations is perpendicular to that in skeletal stretching vibrations of adjacent groups, coupling does not occur. Consequently, the frequency of out-of-plane hydrogen bends *changes markedly/does not change appreciably* in going from one molecule to another with similar double bond structures.

↕H, CH_3, C=C, CH_3—CH_2, H↕

(33)

general direction of skeletal stretching vibrations

A119 α-picoline 740, 760, and possibly 1000 cm^{-1}; acetophenone 690, 760, 960; cyclohexene 710, 880, 920.

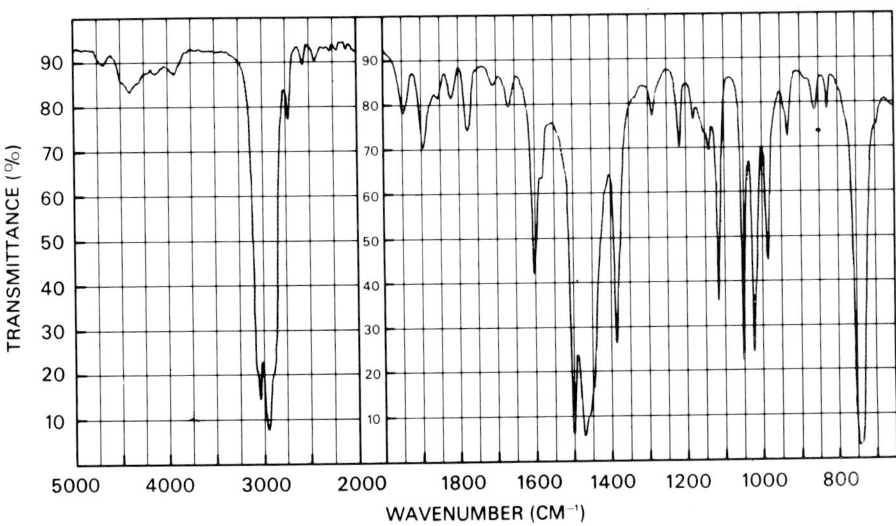

Fig. 3.11. Spectrum of liquid run as a capillary film.

F121 Moreover the precise frequency is often characteristic of substitution on the alkene or aryl group concerned. Use Chart 3 (on the folded sheet in the pocket at the back) to select the compound whose spectrum is given in Fig. 3.11 from the following possibilities:

CH_3, CH_3	CH_3	CH_3, CH_3	CH_3
(34)	**(20)**	**(21)**	**(35)**

A120 does not change appreciably.

F122 Although in most cases we can identify out-of-plane bending vibrations with certainty, interpretation is not always unambiguous. Consult Chart 2 and write down a type of vibration which might be confused with an out-of-plane C—H bend.

A121 (**34**). (**20**) would show two bands of comparable intensity, (**21**) and (**35**) would not show strong bands below 800 cm^{-1}. Substitution correlations like this can be very useful but caution is necessary in cases where the substituents are highly polar. This does not matter in the present case.

F123 Thus, if a benzenoid compound is known to contain chlorine, it is usually difficult to distinguish between the C—Cl stretching band and the out-of-plane benzenoid C—H band with certainty. A similar difficulty may arise with chloroalkenes. Furthermore, if a molecule contains both benzenoid and alkene groups

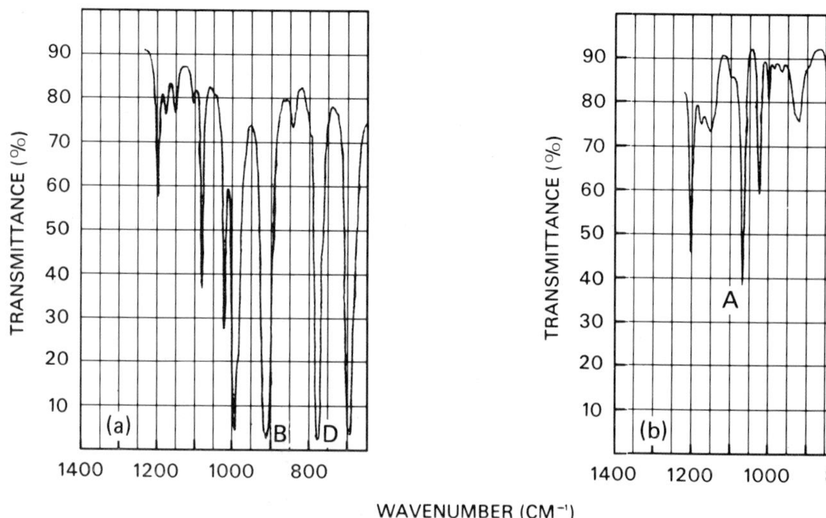

Fig. 3.12. Low frequency region of two spectra of aromatic liquids run as capillary films.

it will not always be possible to distinguish between the benzenoid and alkene out-of-plane C—H bends. You are given (Fig. 3.12) two partial spectra; one of benzyl chloride (**36**) and the other styrene (**37**).

(**36**) (**37**)

Which is which? Which of the labelled bands in each case is a benzenoid out-of-plane bend?

A122 C—Cl, C—Br, C—F, C—S stretches. C—H rocks and wags.

F124 Which of the following statements is/are false?

(a) The frequency of the C—Cl stretching vibration in chlorobenzene is so close to that of CH out-of-plane bends that the two vibrations couple.

(b) The usefulness of absorptions due to benzenoid and alkene out-of-plane bending vibrations comes not only from the fact that they are usually intense and sharp, but also that their frequencies are diagnostic.

(c) It is usually possible to assign vibrations to absorptions below 1400 cm^{-1} involving out-of-plane $=C<_{H}$ bends, C—O stretches and C—Cl stretches.

A123 (a) Styrene (**37**). (b) Benzyl chloride (**36**). The monosubstituted alkene out-of-plane deformations at 910 and 995 cm^{-1} are the key here, although the enormous strength of absorptions below 800 in (b) could be associated with C—Cl. Of those bands labelled, only C and D are due to benzenoid C—H bends. It is very difficult to make assignments within the range 850—600 cm^{-1} for Fig. 3.12 (b).

A124 (a). The vibrations occur at right angles to one another.

Absorption Characteristics of the Carbonyl Group

F125 We have now surveyed, in a rather general way, the distribution of those frequencies in an infrared spectrum which are characteristic of certain functional groups. The position at which these frequencies occur can be rationalized in terms of their dependence on the masses of the atoms and the strengths of the bonds involved in the vibration. Furthermore, we are able to apply the principles given in Part 1 to explain why some vibrations give rise to very reliable frequencies while others do not. We shall now consider in detail one particular vibration, the stretching of the carbonyl group, as an example to illustrate the factors which may influence a vibration's precise frequency in a given molecule. The carbonyl group is important in organic chemistry, and its characteristic frequency has been extensively studied in a wide range of compounds. Which of the following compounds contain a carbonyl group? Look up their structures if necessary.

(a) ethyl alcohol (b) benzoic acid (c) acetyl chloride
(d) picric acid (e) urea (f) cyclopentanone
(g) dioxan (h) methyl salicylate (i) succinamide
(j) acetic anhydride

F126 We see that the carbonyl group occurs in a wide range of organic molecules of very diverse properties, e.g.

R—C(=O)—H	R—C(=O)—R	R—C(=O)—OH	R—C(=O)—OR
aldehyde	ketone	carboxylic acid	carboxylic ester

$$R-C(=O)-Cl \qquad R-C(=O)-NH_2 \qquad R-C(=O)-O-C(=O)-R$$

acid chloride   amide   anhydride

Consult Chart 2 and write down the range over which the C=O stretch of carbonyl compounds can be observed in the infrared. Note whether or not this range is larger or smaller than that in which C=C stretches are found.

A125 (b), (c), (e), (f), (h), (i), (j).

F127 It is indeed fortunate that this range is not seriously overlapped by other characteristic frequency ranges, or its usefulness would be much impaired.

Which functional groups can also produce characteristic frequencies in the range of the carbonyl group frequency?

A126 1620—1860 cm^{-1}, larger.

F128 Although C=C and C=O frequency ranges overlap to a very small extent, there is usually no confusion as to which group is responsible for a band occurring in the 1600—1700 cm^{-1} region. The C=O group is much *more/less* polar than the C=C group so that the intensity of a C=O band is much *less/greater* than that of a C=C band.

A127 C=C, C=N stretch; NH bend; to a small extent in each case.

F129 If we look at the C=O frequency of a series of compounds in any one of the classes listed in **F126**, we find that the range is in fact quite narrow. Once this kind of empirical information has been collated it is possible not only to recognize that a carbonyl group is present in the molecule but to place the compound in one of the above classes according to the position of the carbonyl band within the overall range. This information is summarized in Chart 4.

Which of the following molecules will show:

(a) the highest; (b) the lowest C=O stretching frequency?

CH_3COOEt CH_3CONH_2 CH_3COCH_3 CH_3CHO CH_3COCl $CH_3COOCOCH_3$

A128 more, greater. Recall that the strength of an infrared absorption increases with increasing change of dipole moment during the vibration.

F130 Use the chart to place the following non-conjugated carbonyl compounds in order of increasing frequency: aldehydes, esters, amides, acyclic ketones, acid fluorides, acid chlorides.

A129 (a) $CH_3COOCOCH_3$; (b) CH_3CONH_2.

F131 The first point we notice about the position of carbonyl frequencies within the 'carbonyl region', is the strong dependence on the groups attached to the carbon atom. Consider the general carbonyl compound where X is a more electro-

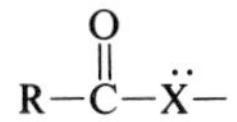

$$R-\overset{\overset{\displaystyle O}{\|}}{C}-\ddot{X}-$$

negative atom than carbon, and bears one or more electron pairs (e.g. $-\ddot{N}-$ (with a third bond below N), $-\ddot{\underset{..}{O}}-$ $:\ddot{\underset{..}{Cl}}-$). The group X influences the strength, length and bond order of the carbon–oxygen bond. These parameters are interdependent. As the bond order increases the bond strength increases and the bond length decreases. Which of the following statements is/are true?

(a) C=C is shorter than C—C.
(b) C≡N is weaker than C—N.
(c) If the group X lowers the bond order of C=O in $R-\overset{\overset{\displaystyle O}{\|}}{C}-X$ it will raise its frequency.

A130 amides, acyclic ketones, aldehydes, esters, acid chlorides and acid fluorides.

F132 The atom X exerts its influence on the carbonyl group in two ways which oppose one another.

(a) a negative inductive effect which withdraws electrons, thereby making the bond shorter.

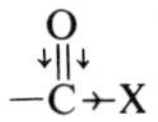

$$-\overset{\overset{\displaystyle O}{\downarrow\|\downarrow}}{C}\leftarrow X$$

(b) a resonance effect which allows structures of type (ii) to contribute significantly to the overall structure of the group (iii).

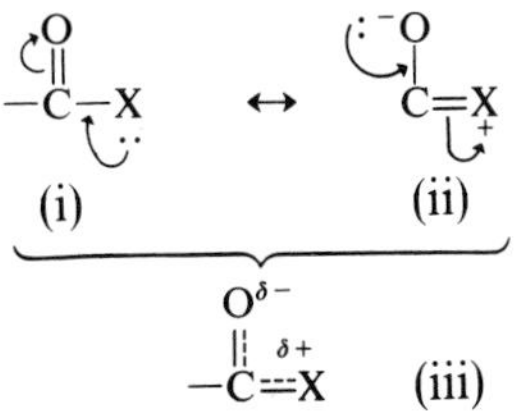

$$\underset{\text{(i)}}{-\overset{\overset{\displaystyle O}{\|}}{C}-\ddot{X}} \leftrightarrow \underset{\text{(ii)}}{\overset{\overset{\displaystyle :\!\bar{O}}{|}}{C}=\overset{+}{X}}$$

$$\overset{\overset{\displaystyle O^{\delta-}}{\|}}{-C}\overset{\delta+}{=\!=}X \quad \text{(iii)}$$

Which of these two effects increases and which decreases the bond order of the carbonyl group?

- - -

A131 (a). (b) is false because a triple bond is of higher order than a single bond. (c) is false since, if the bond order is lowered, the strength and therefore the force constant is lowered. Hence the frequency is lowered.

F133 Select the correct words in the following statement.

The inductive effect in RCOX— causes an *increase/decrease* in carbonyl stretching frequency whereas the resonance effect causes a *decrease/increase.*

- - -

A132 Inductive effect increases.
Resonance effect decreases.

F134 In an earlier frame you found empirically that the carbonyl frequencies of some classes of organic compounds increased in the order, amides, ketones, aldehydes, esters, acid chlorides, acid fluorides. Use this order to deduce which effect appears to be strongest in a carbonyl compound where X = (a), F, (b) Cl, (c) O, (d) N by comparison with the case where X = C (ketone).

- - -

A133 increase, decrease.

F135 Which of the following molecules has the highest carbonyl stretching frequency?

$CH_3(Cl)C{=}O$ $F(Cl)C{=}O$ $H_2N(CH_3)C{=}O$ $Cl(CH_3)C{=}O$ $Cl(OCH_3)C{=}O$ $F(CH_3)C{=}O$

- - -

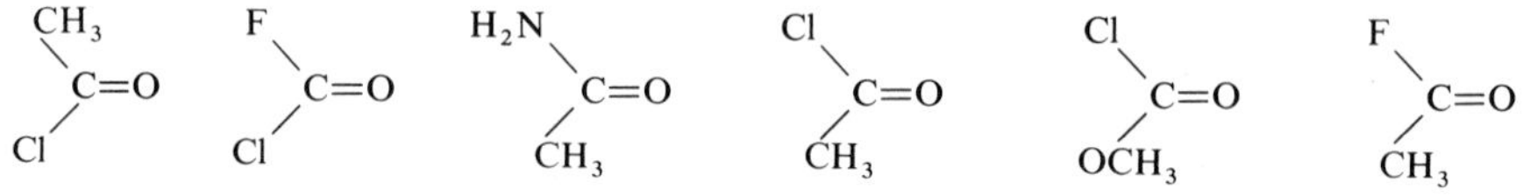

A134 (a), (b), (c) inductive; (d) resonance.

F136 Although the resonance effect exerts the greatest influence only in the case X = N for the group —COX, it is dominant in unsaturated and aromatic carbonyl compounds. Thus cyclohexenone (**47**) and tetralone (**48**) are expected to absorb at *lower/higher* frequency than cyclohexanone (**49**).

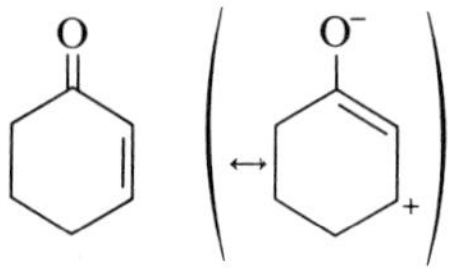

O ↔ O^- ... + **(38)**   O ↔ O^- ... + **(39)**   O **(25)**

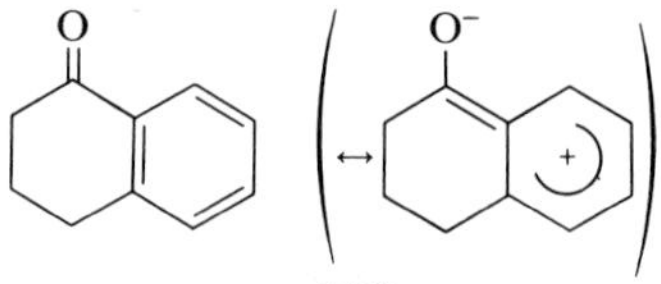

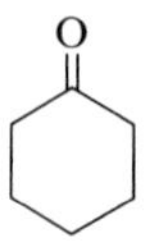

- - -

A135 $FC(Cl)=O$

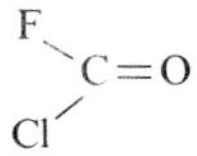

In all but one molecule, acetamide, the inductive effect is dominant and the correct answer involves the largest net inductive withdrawal from the carbonyl group.

F137 The effects of substituents on carbonyl groups are cumulative. Write down the expected order of increasing carbonyl stretching frequency for the three compounds below:

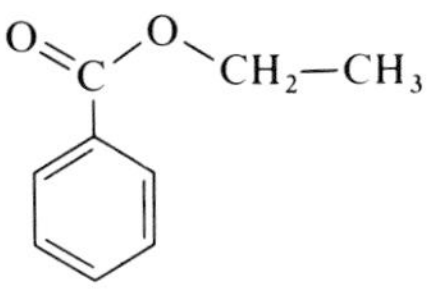

(40) $C_6H_5-CO-O-CH_2-CH_3$ (41) $C_6H_5-CO-CH_3$ (42) $C_6H_{11}-CO-O-CH_2-CH_3$

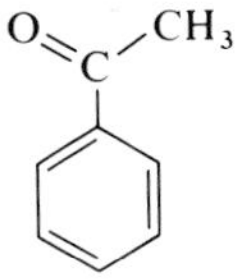

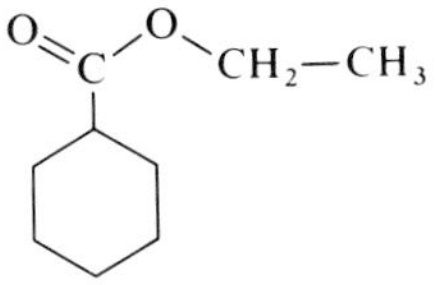

A136 lower. Contributions to the structure from the canonical forms in the brackets lower the bond order.

F138 The correlation of frequencies with inductive and resonance effects that we have been using accounts qualitatively for the observed trends. It is necessary however to exercise some caution since we have assumed a very simple picture of the carbonyl vibration. The carbonyl vibration is not located entirely within the carbonyl bond but involves some motion of other atoms in the molecule. Changes in the composition and structure of the rest of the molecule will therefore exert a mechanical influence on the 'carbonyl frequency' which will vary from one molecule to another. In the compounds we have already studied the mechanical effect is negligible. In the case of cyclic ketones on the other hand, it is this mechanical effect which is primarily responsible for the observed increase in the carbonyl frequency as the ring size decreases below six atoms. In large ring and open chain ketones the movement of the carbon atom as the C=O bond stretches is taken up by small displacements of the adjacent carbon atoms as shown in (I). This is not possible with the rigid planar 3-, 4- and 5-membered rings and the stretching of the C=O bond results in a compression of the adjacent C—C bonds (II). The resistance to the C=O stretching results in a progressive rise in the frequency of this vibration as the ring size decreases.

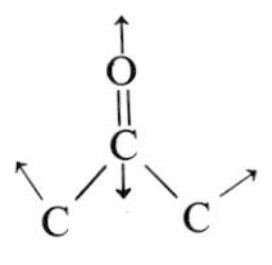

(I) bending effect when C=O stretches in large rings

(II) compression effect when C=O stretches in small rings

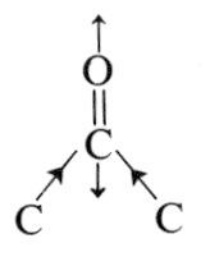

The carbonyl group in **(43)** is contained within a five-membered ring, yet its stretching frequency is found to be near 1785 cm^{-1}. What does this tell us about the C—CO—C bond angle in **(43)**?

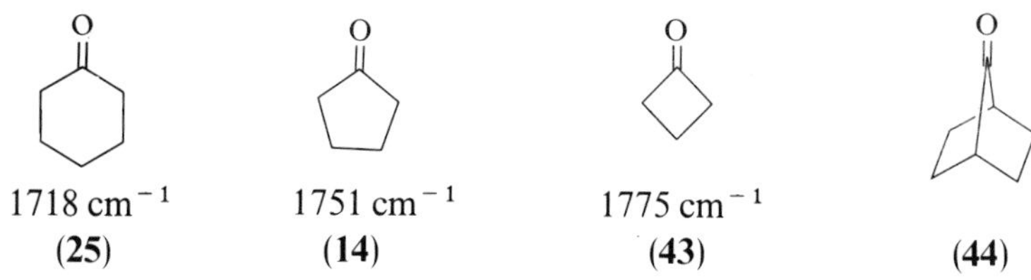

A137 **(41)** < **(40)** < **(42)**. **(40)** will be lower than **(42)** because **(40)** is a conjugated ester. **(41)** will be lower than **(40)** because **(41)** is a ketone and **(40)** is an ester.

F139 Using Chart 4 and accumulating the effects mentioned so far, deduce which of the following three compounds would give a spectrum showing strong absorption at 1720 cm^{-1}.

(38) **(45)** **(46)**

A138 The bond angle in **(44)** is close to that in cyclobutanone, i.e., about 90°.

F140 Assign the most appropriate frequencies for these compounds from the following: 1775, 1750, 1745, 1700 cm^{-1}.

(47) **(48)** **(49)** **(50)**

A139 **(45)**

Saturated six-membered ring ketones absorb at the same frequency as open chain ketones. Chart 4 shows that αβ-unsaturation in ketones causes a drop in the carbonyl frequency of 20—40 cm^{-1}.

F141 Which of the following statements is/are false?

(a) There are many carbonyl compounds which show only weak absorption between 1600 and 1900 cm^{-1}.

(b) Increase in the angle ϕ in the molecular fragment leads to an increase in the carbonyl stretching frequency.

(c) The order of carbonyl stretching frequency given below is a result of the resonance effect causing a polarity in (**51**) which is opposite to that in (**38**).

(**51**) > (**25**) > (**38**)

A140 (**47**) 1750; (**48**) 1775; (**49**) 1700; (**50**) 1745. The heteroatoms in (**47**) and (**50**) are not directly attached to the carbonyl group and hence the stretching frequencies are close to that of cyclopentanone. (**47**) is slightly higher than (**50**) because oxygen is more electronegative than nitrogen. The values for (**48**) and (**49**) follow by analogy from open chain esters, ketones and amides.

A141 (a) and (c). Weak carbonyl absorptions are unknown. Resonance stabilization of carbonyl always operates by electron transfer towards the oxygen atom. The high frequency of the lactone (**51**) is due to the inductive effect of the σ-bonded oxygen atom which operates in the opposite direction.

The Elucidation of Structure from Infrared Spectra

Example 1

F142 Now let us attempt to determine the structure of an unknown organic compound using its infrared spectrum and a knowledge of its molecular formula. Figure 3.13 is the spectrum of a colourless liquid with molecular formula C_8H_7N.

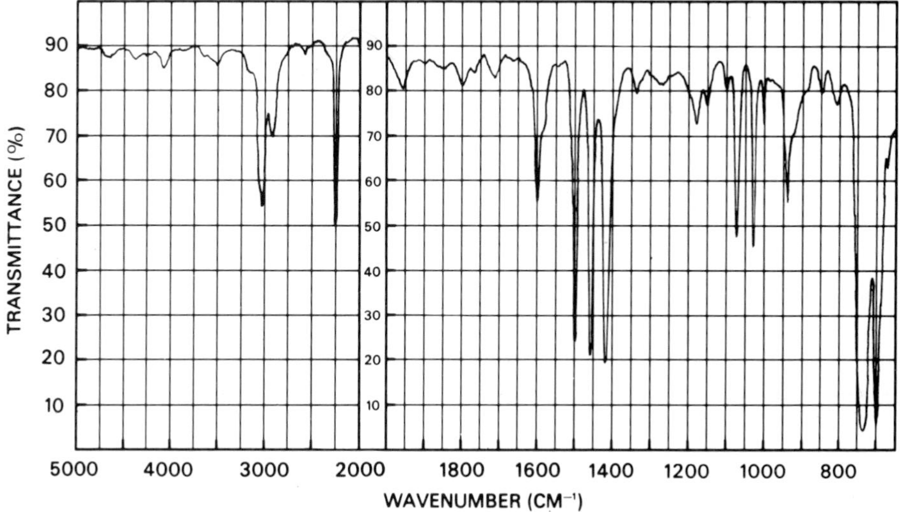

Fig. 3.13. Spectrum of a liquid of formula C_8H_7N run as a capillary film.

We look first at the region above 1500 cm^{-1} where, you recall, the most reliable diagnostic absorptions are to be found. Write down the frequencies of the principal bands in this region (ignoring those showing an absorption of less than 20 %).

F143 Which of these bands corresponds to stretching vibration of bonds by which hydrogen is attached to other atoms?

- - -

A142 1600, 2250, 2900, 3050 cm^{-1}.

F144 The band at 2250 cm^{-1} is probably associated with a *double/triple* bond while the band at 1600 cm^{-1} is associated with a *double/triple* bond.

- - -

A143 2900, 3050 cm^{-1}.

F145 Take into account the molecular formula and the moderate intensity of the 2250 cm^{-1} band, and decide which of the following groups having frequencies in this region is likely to be responsible for the absorption in the spectrum.

$-C{\equiv}C-$ $\quad -C{\equiv}N$ $\quad N{\equiv}\overset{+}{N}-$ $\quad -N{=}C{=}O$ $\quad -\overset{+}{N}{\equiv}\overset{-}{C}$

- - -

A144 triple, double. Some cumulated double bonds, e.g. N=C=O, also give bands over 2000 cm^{-1}.

F146 Use Chart 2 to eliminate one of these two groups.

- - -

A145 $-C{\equiv}N$, $-\overset{+}{N}{\equiv}\overset{-}{C}$. C≡C would not normally give as strong a band as that in Fig. 3.13.

F147 If the 2250 cm^{-1} absorption is correctly identified as $-C{\equiv}N$, we have accounted for the nitrogen atom in the formula. Return now to the two bands near 3000 cm^{-1}, and using Chart 2, suggest groups which may be responsible in each case.

- - -

A146 $-\overset{+}{N}{\equiv}\overset{-}{C}$. Isocyanides absorb below 2200 cm^{-1}.

F148 The presence of benzenoid C—H bonds in a molecule would imply the existence of other vibrations characteristic of a benzenoid ring. Is the spectrum consistent with the presence of a benzenoid ring?

- - -

A147 2900 cm^{-1} aliphatic CH.
3050 cm^{-1} benzenoid or alkene CH.

F149 Write down the frequencies of the aromatic bands present in the spectrum due to ring stretching vibrations and out-of-plane C—H bending vibrations.

A148 Yes. Absorption bands are to be found in the three other 'benzenoid' regions given in Chart 2.

F150 The presence of a benzenoid ring in the molecule accounts for six carbon atoms. Write down as many structural formulae as you can which are consistent with the molecular formula, and the presence of aliphatic CH and CN.

A149 1600, 1500 cm^{-1} (ring stretching); 740, 700 cm^{-1} (out-of-plane C—H bending).

F151 We now have to choose between these four. Which of the following regions of the spectrum will allow us to do this?

600—900 cm^{-1}, 900—1400 cm^{-1}, 1400—3000 cm^{-1}

A150 $C_6H_5CH_2CN$; CH_3 / CN (ortho); CH_3 / CN (meta); CH_3 / CN (para)

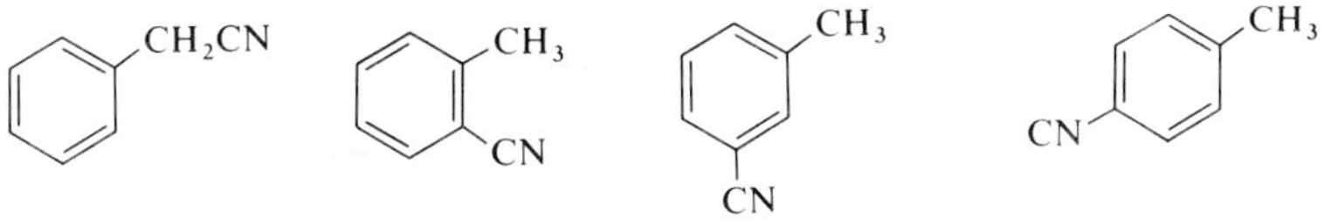

F152 Using the information in Chart 3 identify the unknown molecule.

A151 600—900 cm^{-1}. This is the region within which the out-of-plane bending vibrations absorb, and these are characteristic for benzenoid substitution patterns (**F121**).

Example 2

F153 The spectrum shown in Fig. 3.14 is that of a crystalline solid, C_9H_8O. The single oxygen atom must be present as either

(a) a ketonic/aldehydic carbonyl group;
(b) an ether linkage, C—O—C;
(c) a hydroxy' group, C—O—H.

Which of these three is supported by the spectrum?

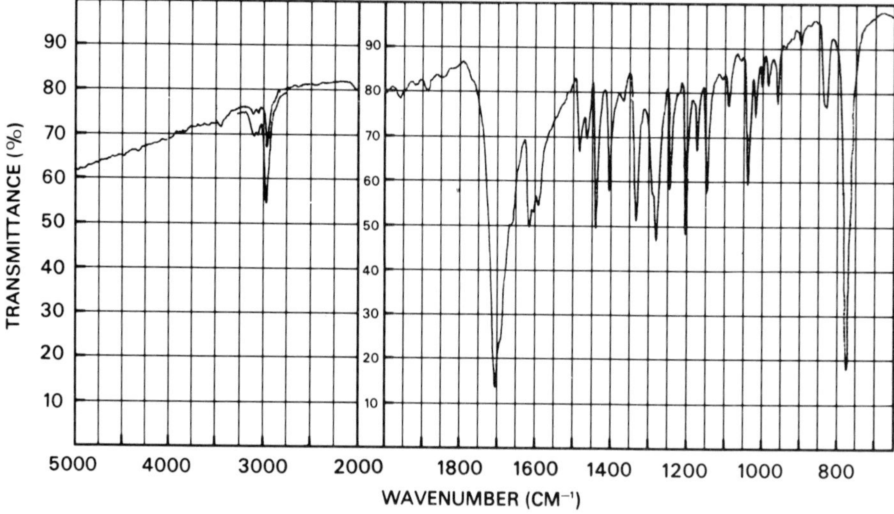

Fig. 3.14. Spectrum of a crystalline solid of formula C_9H_8O run as a KBr disc. (The C—H region has been re-run using a second disc containing more of the sample. Incidentally, the sloping background here is due to inadequate grinding of the sample.)

A152 CH_2CN The two strong bands at 700 cm^{-1} and 740 cm^{-1} are characteristic of monosubstituted benzenoid rings.

F154 The distinction between aldehyde and ketone is unambiguous in this case. Which do we have, and what vibration is present or absent, allowing you to make the choice? (Study Chart 2).

A153 (a). The spectrum contains a strong carbonyl absorption.

F155 Thus you have established the presence of a ketonic carbonyl group. The possibility of a benzenoid ring is always considered whenever a molecular formula contains six or more carbon atoms.

Does the spectrum contain bands consistent with the presence of a benzenoid ring?

A154 Ketone. There is no aldehydic C—H stretching band near 2700 cm^{-1}.

F156 Select three frequencies from the following list taken from the spectrum which are consistent with the presence of a benzenoid ring: ~3150, 2960, 1700, 1610, 1480, 1440, 1335, 960, 775 cm^{-1}.

A155 Yes. Relevant absorption is to be found in the 'benzenoid' regions in Chart 2.

F157 An infrared spectrum can often tell us the substitution pattern of a benzenoid ring. Which of the absorptions listed in **A156** would you use for this purpose?

A156 Absorption above 3000 cm^{-1}, 1610, 1480, 775 cm^{-1} (any three).

F158 What are your conclusions with regard to substitution pattern in the present case?

A157 775 cm^{-1}.

F159 Let us assume that our compound is a 1,2-disubstituted derivative of benzene.

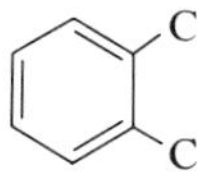

C

C

This stucture accounts for all but one carbon atom, four hydrogen atoms and one oxygen atom. Draw the only *two* possible complete structures consistent with the molecular formula, a 1,2-disubstituted benzenoid ring and a ketonic carbonyl group.

A158 1,2-disubstituted, possibly 1,2,3-trisubstituted (Chart 3).

F160 Which band in the spectrum enables us to distinguish between these two structures and which one is correct?

A159

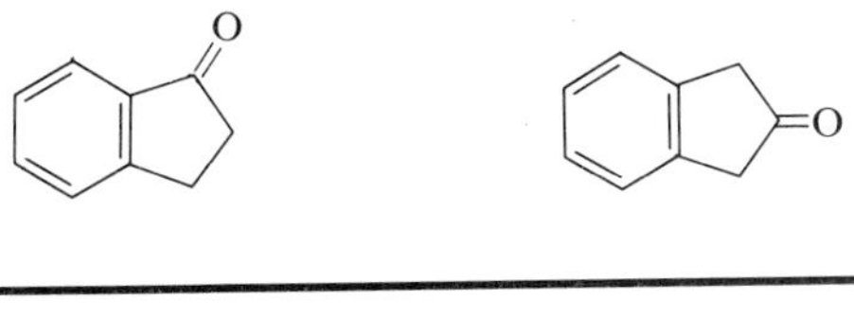

O

O

Example 3

F161 Consider now the spectrum in Fig. 3.15, which was obtained from a liquid of molecular formula $C_4H_6O_2$. List the bands above 1500 cm^{-1} and suggest which functional group is responsible in each case.

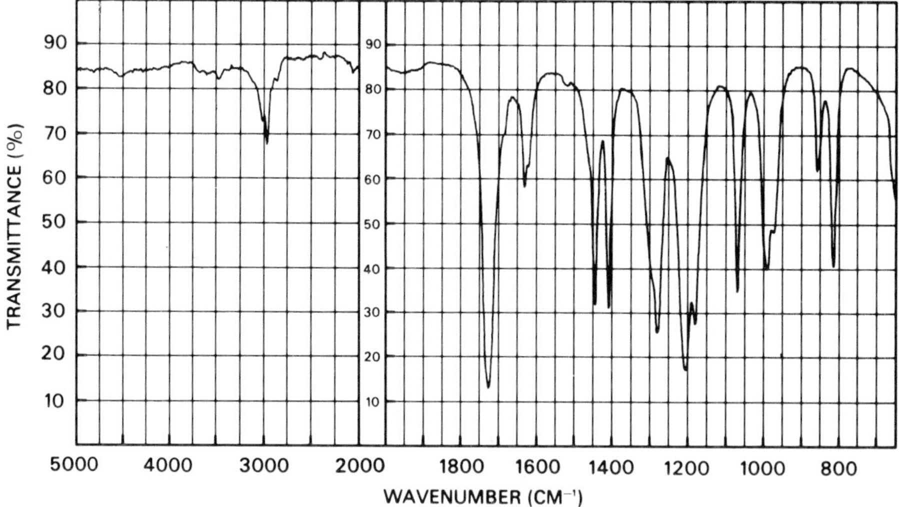

Fig. 3.15. Spectrum of a liquid of formula $C_4H_6O_2$ run as a capillary film.

A160 The carbonyl band. Its frequency indicates that the C=O group in the 5-membered ring is conjugated with the benzenoid ring.

F162 What type(s) of carbonyl group could be responsible for the band at 1725 cm^{-1}?

A161 aliphatic C—H 2950 cm^{-1}
C=O 1725 cm^{-1}
C=C 1635 cm^{-1}

You may have included hydrogen-bonded —OH in your answer. These bands are always very broad and there is never any confusion between them and the much sharper C—H bands which occur near 3000 cm^{-1} in the spectra of all organic compounds. O—H bands will be discussed later.

F163 Notice that the 1635 cm^{-1} band also occurs in the carbonyl range. Why can we confidently assign this to C=C?

A162 aldehyde, acid, ketone; $\alpha\beta$-unsaturated ester, α-diketone.

The molecular formula excludes other possibilities.

F164 Eliminate or confirm the presence of the aldehyde group.

A163 The 1635 cm^{-1} absorption is far too weak to be due to C=O.

F165 Is the compound a carboxylic acid?

A164 Aldehydic C—H gives rise to absorption near 2700 cm^{-1}. This is not present and so the compound is not an aldehyde.

F166 You are thus left with the choice of a ketone or an $\alpha\beta$-unsaturated ester. The presence of C=C has already been established (**F163**). However, you can further distinguish between these two possibilities by seeking evidence for a bond which is present in an ester and absent in a ketone. To which bond are we referring?

A165 No. The OH group of carboxylic acids give rise to a broad band near 3000 cm^{-1} (Chart 2 and **A161**).

F167 You have now tentatively established that the compound is an $\alpha\beta$-unsaturated ester from the position of its carbonyl frequency and the absence of certain other characteristic vibrations. The presence of C=C has already been noted (1635 cm^{-1}). Write down as many structural formulae as possible for the compound on the basis of its infrared spectrum and molecular formula.

A166 C—O is present only in an ester and there is strong evidence for it in the form of an intense band at 1210. Thus, the ketone is tentatively eliminated. (The possibility of ketone and ether functional groups occurring separately is not ruled out yet.)

F168 Compare the spectrum with information given in Chart 3. Is the C—H out-of-plane bending region (600—1000 cm^{-1}) consistent with the presence of the vinyl group?

A167 $H_2C{=}CH{-}C({=}O){-}O{-}CH_3$ is the only possibility.

The C=C and C=O account for three carbon atoms and one oxygen atom. Since there is only one more carbon atom in the molecule the alkyl group of the ester can only be methyl. Vinyl acetate, $CH_2{=}CH{-}O{-}CO{-}CH_3$, is excluded because its carbonyl group is not conjugated. A compound containing separate ketone and ether functions together with a double bond would be impossible with just four carbon atoms.

A168 Not entirely. According to the chart, a vinyl group should give rise to strong bands at about 910 and 990 cm^{-1}. There is certainly a strong band at 990 cm^{-1} but nothing at 910. The evidence is therefore somewhat inconclusive, but, of the possibilities, the spectrum is more consistent with vinyl than any other alkene structure. The reliability of C—H out-of-plane bands is much impaired by the presence of highly polar substituents on the double bond. In this simple molecule, however, the structure determination is quite definite and the evidence of the out-of-plane bands is not essential to the argument.

Revision Summary

Functional groups can be recognized by their vibrational frequencies manifested as characteristic absorption bands in infrared spectra. The frequencies at which absorption occurs for all the common functional groups are now well known and so analysis of a spectrum can often be carried out using empirical information only (e.g. Chart 2). One can go further than this, however, and explain, in a qualitative way, the distribution of stretching frequencies by recalling the stretched spring analogy.

Stretching frequency depends on the strength of the bond and the masses of the atoms at either end. Thus all X—H bonds have much higher frequency than X—Y bonds, where X and Y are carbon, nitrogen or oxygen. Similarly X—Y, X=Y and X≡Y increase in frequency in the order given. (see Chart 2).

The fact that C—O, C—N, N—N stretching and some CH bending vibrations often have similar frequencies leads to coupling between them and gives rise to **skeletal vibrations.** The corresponding absorptions, which occur below 1500 cm^{-1}, cannot be ascribed to a particular small group and they are not very useful for characterization purposes. Exceptions to this are C—O and C—Cl. Although its frequency is not particularly reliable, the exceptional strength of bands due to the C—O stretching vibration allows it to be recognized readily within the skeletal region. Similarly, absorption due to C—Cl stretching is always very strong and its frequency is confined to narrower limits than most other X—Y stretches.

Coupling between vibrations usually occurs only when they

(i) have very similar frequencies;
(ii) occur in adjacent bonds;
(iii) involve motions in a common direction.

Skeletal vibrations are responsible for most of the bands in the finger-print region (1400—900 cm^{-1}), which, in the spectra of most organic compounds, is a complex area of absorption, whose overall pattern usually identifies a particular structure.

The most useful bending vibrations are those involving out-of-plane CH bending motions in unsaturated compounds. These give rise to strong absorptions between 1000 and 600 cm^{-1}, and the precise frequencies are characteristic of substitution patterns of alkenes and benzenoid rings (Chart 3).

The C=O group is very strongly absorbing in the infrared, and its frequency is very characteristic of the type of carbonyl compound being measured. The relative positions for various types of compounds within the overall carbonyl region can be understood in terms of resonance and inductive effects. Electron donation by resonance leads to greater single-bond character in the C=O bond and hence leads to a lower frequency for the corresponding stretching vibration. Inductive electron withdrawal results in a higher C=O bond order and hence higher frequency. Thus conjugation leads to a lowering of carbonyl frequencies whereas compounds in which a negative inductive effect is dominant (such as the acid halides) have high frequencies. The frequency of C=O stretching also increases as the C—CO—C bond angle decreases.

Now try the multiple choice test at the beginning of this part. Write your answers in a vertical column and mark your script as before. Then note the remarks below.

Less than 6 Consult your instructor as to whether you should go on or seek extra tuition. If you are using the programme for private study, re-read the frames pertinent to the questions you answered incorrectly and, if possible, consult the references for further reading. When you are satisfied you fully understand the material, read the paragraph at the end of this Part and start the next Part.

More than 6 Revise the material pertinent to those questions you answered incorrectly. If your instructor does not want you to consult the further reading section or attempt any of the questions which follow, read the paragraph at the end of this Part and continue with the next Part.

Questions

1. Compare the conditions which must be fulfilled for significant coupling to occur between two vibrations, with those necessary for Fermi resonance. In which way are they similar? Point out the differences.

2. How would you use infrared spectroscopy to distinguish between each of the following pairs of organic compounds? Name the vibrations and approximate positions of absorptions in each case.

(i) CH_3CH_2COCl and $ClCH_2CH_2OCH_3$

(ii) cyclopentanone (H_2C–CH_2–C=O–CH_2–H_2C ring) and 3-methylcyclobutanone (O=C–CH_2–CH(CH_3)–H_2C ring)

(iii) C_6H_5–$CH_2CH_2CH_3$ and C_6H_5–CH_2–CH=CH_2

(iv) 1,2,4,5-tetramethylbenzene (CH_3 ×4) and hexamethylbenzene (CH_3 ×6)

(v) $CH_3CH_2CH_2COOH$ and

$$\begin{array}{l} \qquad\quad OH \\ H_2C-CH \\ H_2C-C-CH_3 \\ \qquad\quad OH \end{array}$$

(vi) $C_6H_5C{\equiv}CH$ and $C_6H_5CH{=}CH_2$

(vii) CH_3CH_2CHO and CH_3COCH_3

(viii) $(CH_3)_3N$ and $CH_3CH_2CH_2NH_2$

3. Write down the structure of the compound C_3H_6O whose infrared spectrum is shown in Fig. 3.16.

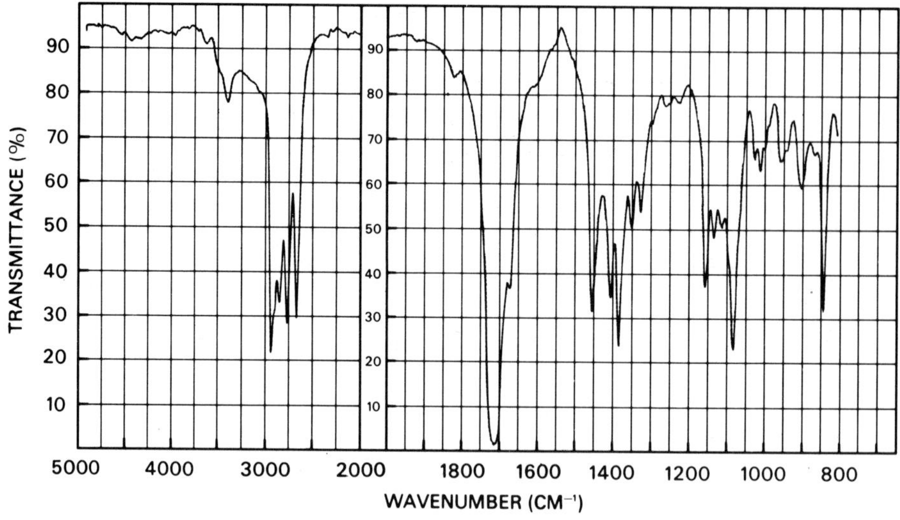

Fig. 3.16. Spectrum of a liquid of formula C_3H_6O.

4. A compound of molecular formula C_4H_8 has infrared absorption near 3100, 1645, 1420, 995 and 915 cm^{-1}. Suggest a structure which will be consistent with this data.

5. The infrared spectra of six organic compounds are shown in Fig. 3.17. Assign each spectrum to one of the following structures, giving reasons for your choice.

Cl (on a benzene ring) — O=C–OCH_2CH_3 (on a benzene ring) — (cyclohexene)

NH_2 and CH_3 (para on a benzene ring) — $\underset{CH_2-CH_2}{OCH_3\ \ OCH_3}$ — CH_2–C(=O)–OH (on a benzene ring)

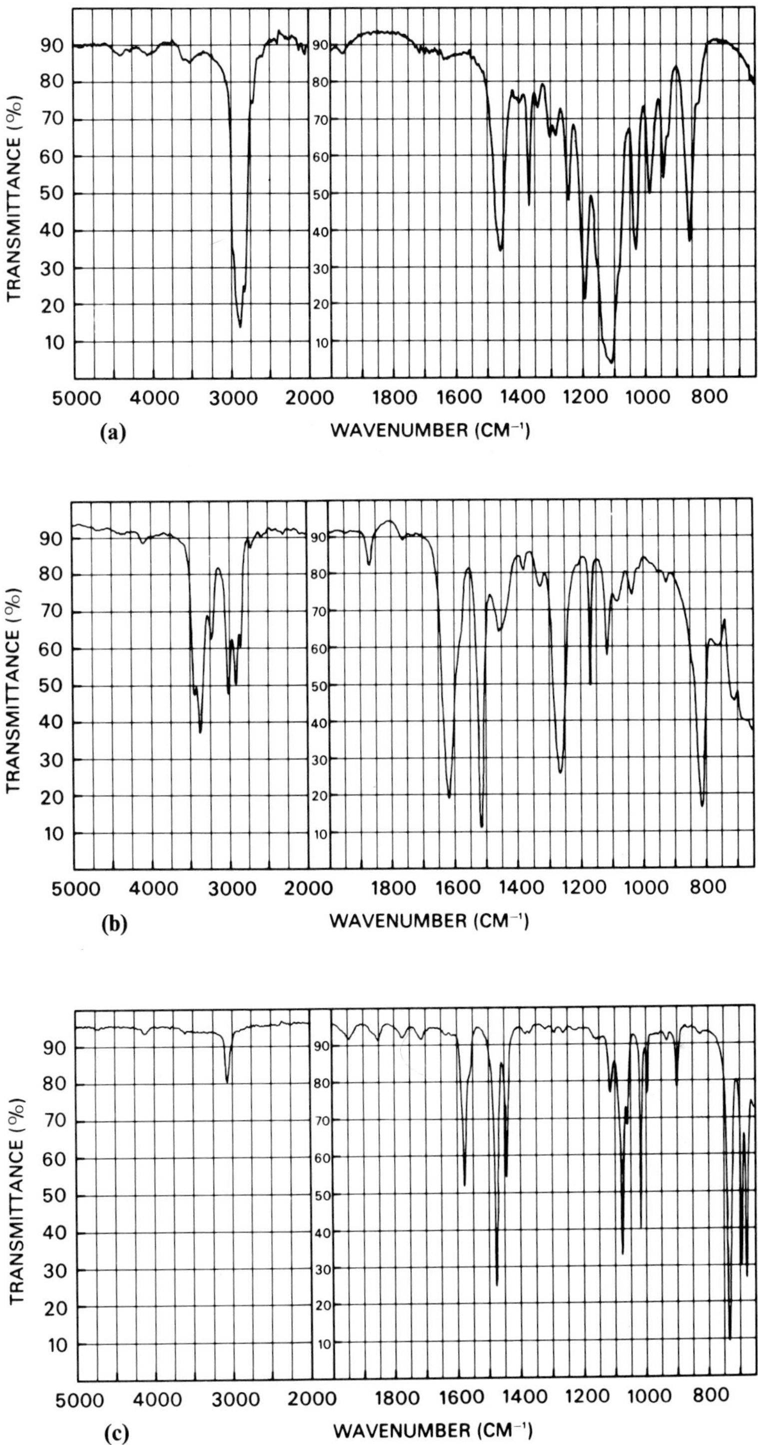

TRANSMITTANCE (%)
90
80
70
60
50
40
30
20
10
5000
4000
3000
2000
1800
1600
1400
1200
1000
800
(a)
WAVENUMBER (CM⁻¹)
(b)
(c)

Fig. 3.17

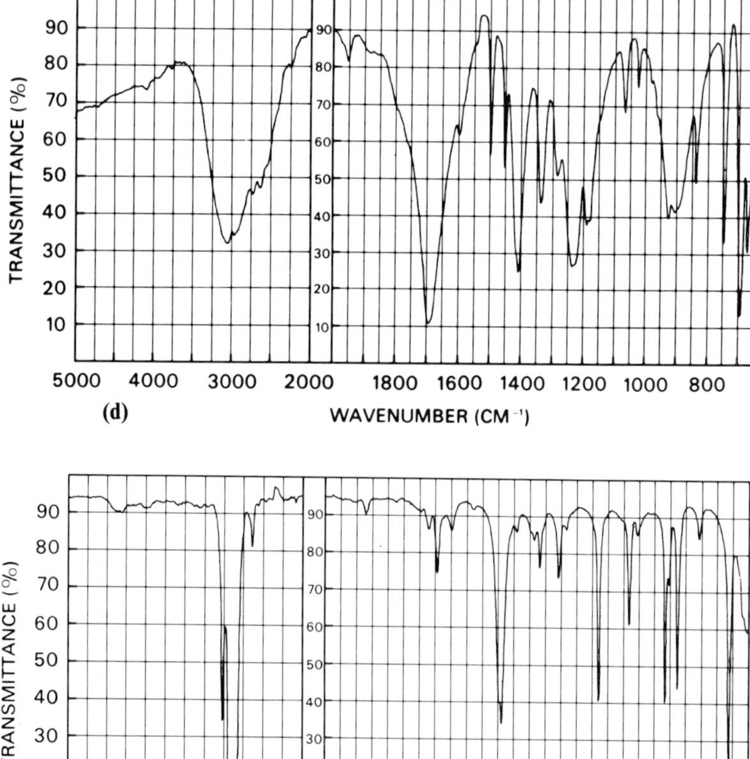

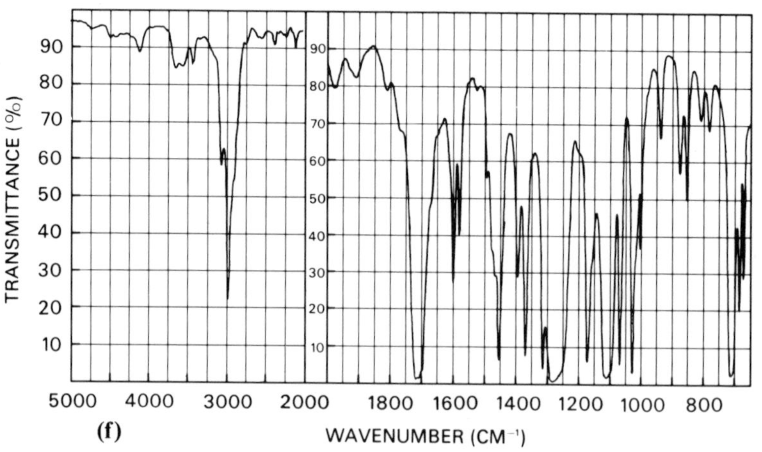

Fig. 3.17—*contd.*

Further Reading

J. R. Dyer, *Applications of Absorption Spectroscopy of Organic Compounds*, Prentice Hall, Englewood Cliffs, N.J., 1965, Chapter 3.

R. M. Silverstein and G. C. Bassler, *Spectrometric Identification of Organic Compounds*, John Wiley & Sons, New York, 1967.

D. H. Williams and I. Fleming, *Spectroscopic Methods in Organic Chemistry*, McGraw-Hill, New York, 1966, Chapter 3.

K. Nakanishi, *Infrared Absorption Spectroscopy*, Holden Day, New York, 1962.

C. N. Rao, *Chemical Application of Infrared Spectroscopy*, Academic Press, New York, 1964.

Mansel-Davies, *Infrared Spectroscopy and Molecular Structure*, Elsevier, Amsterdam, 1963.

A. D. Cross and R. A. Jones, *An Introduction to Practical Infrared Spectroscopy*, Butterworths, London, 1969.

M. Avram and Gh. D. Mateescu, *Infrared Spectroscopy, Applications in Organic Chemistry*, Wiley-Interscience, New York, 1972.

S. F. Dyke, A. J. Floyd, M. Sainsbury and R. S. Theobald, *Organic Spectroscopy—an Introduction*, Penguin, London, 1971.

F. Scheinmann, *Introduction to Spectroscopic Methods for the Identification of Organic Compounds*, Volume 1, Pergamon Press, Oxford, 1970.

You should now be able to extract a considerable amount of information about an unknown compound from its infrared spectrum. Your main aid is empirical data in the form of tables of characteristic frequencies, but your understanding of the reasons for the distribution of these frequencies among a wide variety of vibrations constitutes a most valuable supplement. Indeed, such knowledge is vital if you are to make the most intelligent use of the tables. The speed at and extent to which you will be able to analyse the structures of organic compounds from their spectra now depends on experience and we hope Parts 4 and 5 will help to minimise the time needed to reach a level of competence in spectral interpretation which is adequate for most structural problems.

PART 4

INFRARED ABSORPTION CHARACTERISTICS OF THE COMMON CLASSES OF ORGANIC COMPOUNDS

Aim

Parts 1 to 3 have brought the student to the point where he can use infrared spectra empirically in structure analysis, and in particular, where he can discriminate between significant and relatively unimportant absorption features. He should also be able to rationalize in general terms any observed modifications to normal band frequencies and intensities produced by special structural features in a molecule. The aim of the present section is to demonstrate how this knowledge and understanding can be applied to the infrared absorption characteristics of each class of organic compound. To a large extent, the student will be able to deduce for himself what the main absorption characteristics of a class of compounds will be by using concepts he has already met in previous parts of the programme.

New Terms and Concepts

Cumulated and conjugated double bonds; the effects of hydrogen bonding on O—H, N—H and C=O stretching frequencies; carboxylic acid dimers.

Prior Knowledge

The material covered by Parts 1 and 3 (see summaries); a superficial knowledge of hydrogen bonding; resonance.

Objectives for Part 4

When you have completed Part 4 you should be able to:

1. State the most reliable infrared absorption characteristics for a given (common) functional group.
2. Predict whether or not certain features of the infrared spectrum of a compound of known structure will vary significantly with the physical state of the sample or its concentration in solution and identify those features.
3. Predict the effect of electronic interaction between functional groups on relevant features of an infrared spectrum.
4. Identify situations where hydrogen bonding is responsible for frequency shifts from normal positions.
5. Use an infrared spectrum, correlation charts and knowledge of concepts and principles given in the programme to choose between plausible structural alternatives for an unknown organic compound.

PART 4: INFRARED ABSORPTION CHARACTERISTICS OF THE COMMON CLASSES OF ORGANIC COMPOUNDS

Introduction

So far we have discussed organic infrared spectroscopy by classifying the absorptions according to vibrational type. You found how convenient this was in using empirical information for the interpretation of spectra. Frequently, however, it is more relevant to be aware of the absorption characteristics of a particular class of compound than of a particular class of vibration. Part 4 presents an opportunity for you to apply your knowledge and understanding of infrared spectroscopy to the prediction of important absorption features of some common types of organic compound. The factual knowledge acquired in this exercise will be valuable and the whole section will constitute a useful (though by no means comprehensive) survey for reference purposes. In many ways Part 4 develops the concepts discussed in Part 3 and many of the questions cannot be answered using only the information given in the preceding frame. In some cases an incorrect response will require reference to previous parts for rectification.

Try the multiple choice test before studying the programme and mark it in the usual way.

Multiple Choice Test

Attempt these questions without reference to the Charts.

1. The presence of C—H bonds adjacent to a C=C bond can be detected by infrared spectroscopy. Select the best order of usefulness and reliability of vibrations for this purpose from the following sequences:

   (a) CH out-of-plane bend > CH stretch > CH in-plane bend > C=C stretch;

   (b) CH stretch > CH in-plane bend > CH out-of-plane bend > C=C stretch;

   (c) C=C stretch > CH stretch > CH out-of-plane bend > CH in-plane bend.

2. Which of the following three classes of compounds give solution spectra which show the least dependence on concentration?

   (a) alcohols;
   (b) aromatic nitro-compounds;
   (c) amides.

3. The most diagnostic feature in the spectra of aldehydes is

   (a) the frequency of the C=O stretching band;
   (b) the fact that the C=O band is a doublet;
   (c) a doublet near 2700 cm^{-1}.

4. Which of the following observations would enable you most readily to distinguish between (**A**) and (**B**)?

(**A**) $C_6H_{11}-C(=O)-OH$ (**B**) 3-hydroxycyclohexanone (ring C=O, OH)

(a) the shape of the —OH stretching absorption;
(b) the frequency of the C=O stretching absorption;
(c) the strength of the C=O stretching absorption.

5. The most characteristic feature in the infrared spectrum of a carboxylic anhydride is

(a) a single carbonyl absorption of exceptionally high intensity;
(b) a pair of absorptions due to the carbonyl group of unusually low intensity;
(c) a pair of carbonyl stretching absorptions at least one of which is very strong.

6. Which of the following bands would be of **most** value in distinguishing between primary and secondary aromatic amides if the spectra were obtained of a dilute solution of the sample in a non-polar solvent?

(a) N—H stretching; (b) N—H bending; (c) C=O stretching.

7. The characteristic frequency of the isocyanate group occurs in the same region as

(a) the C≡N stretching frequency;
(b) CH deformation frequencies;
(c) the C=C stretching frequency.

8. Which of the following compounds would **not** give rise to two absorption bands that could be described as C=C stretching?

(a) $(CH_3)(H)C=C(H)-C(H)=CH_2$ (b) $(H)(CH_3)C=C=C(CH_3)_2$

(c) $(CH_3)_2C=CH-CH_2-HC=C(CH_3)_2$

9. Which of the following compounds might be expected to show the lowest carbonyl frequency and a **broad** absorption band near 3000 cm^{-1}?

(a) $CH_3CO(CH_2)_2COCH_3$

(b) $CH_3COCH_2COCH_3$

(c) $CH_3COCOCH_3$

10. The structural feature responsible for the low frequency of the carbonyl stretch of simple carboxylic acids relative to, say, esters is:

(a)
```
    O       O     O       O     O       O
 \ //  ··  / \  //  ··  / \  //  ··  / \  //
  C      H     C      H     C      H     C
  |            |            |            |
  R            R            R            R
```

(b)
```
        O
      //  ··
R—C        H
      \   /
        O
```

(c)
```
      O···H—O
    //        \
R—C            C—R
    \        //
      O—H···O
```

Comments on marks

Less than 3 Proceed directly with Part 4.

3 to 7 You either have some previous factual knowledge of the contents of Part 4 or you are able to apply the principles learnt in Parts 1 and 2 very effectively. There is little doubt, however, that you would improve your ability to analyse infrared spectra by working through the present section.

More than 7 You can by-pass Part 4 and go directly to the worked examples of Part 5. However, you should consult those frames relevant to any incorrectly answered questions.

Saturated Hydrocarbons

F169 The simplest class of organic compounds is the saturated hydrocarbons. Molecular vibrations are restricted to C—H and C—C stretches, and HCH, CCH and CCC bends. Which of the two bonds, C—H and C—C has the higher stretching frequency?

F170 C—C stretching bands occur in the skeletal region of the spectrum. They are, accordingly, *very useful/of little value* for diagnostic purposes.

- - - - - - - - - - - - - - - - - - - -

A169 C—H.

F171 Which of the two types of bending vibrations, CH or CC, give rise to the more useful absorption bands?

- - - - - - - - - - - - - - - - - - - -

A170 of little value. They are also very weak.

F172 We shall find, therefore, two main groups of useful absorption bands in the infrared spectrum of a saturated hydrocarbon. Which of the following vibrations are responsible?

(a) C—H and C—C stretching;
(b) C—C and C—H bending;
(c) C—H bending and stretching.

A171 CH. (You should recall that CCC bending vibrations have frequencies below the normal range of routine spectrophotometers.) However, some types of CH bending occur in the skeletal region and are of little value.

F173 Use Chart 2 to write down the frequency ranges corresponding to CH stretching and bending in saturated hydrocarbons.

A172 (c).

F174 Let us consider first the stretching vibrations. The CH bonds must be present in one of three groups:

$H_3C—$	$—CH_2—$	$\rangle C—H$
Methyl	Methylene	Methine
	Open-chain or cyclic	

Write down the nomenclature which describes the different stretching vibrations of the methylene group.

A173 2800—3000 cm^{-1}; 1320—1480 cm^{-1}, 690—1215 cm^{-1}

F175 The stretching frequencies of the methylene group are generally the same in cyclic compounds as in open chain systems. The values are 2926 and 2853 cm^{-1} and these rarely vary by more than 10 cm^{-1} except in the case of small ring compounds. Cyclopropane C—H stretching vibrations are found in the region 2990—3100 cm^{-1}.

A bicyclic chloroketone is found to have no absorption between 2000 and 2900 cm^{-1}. Which of the following structures is most likely for the compound?

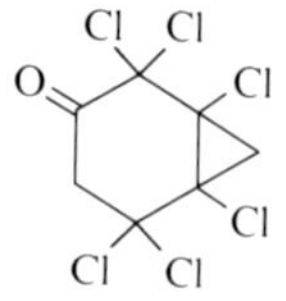

(1) 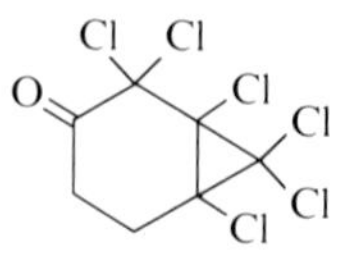 (2) 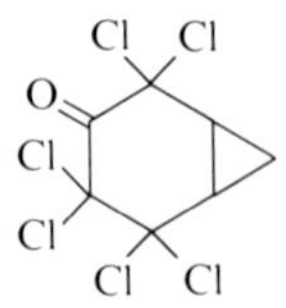 (3)

A174

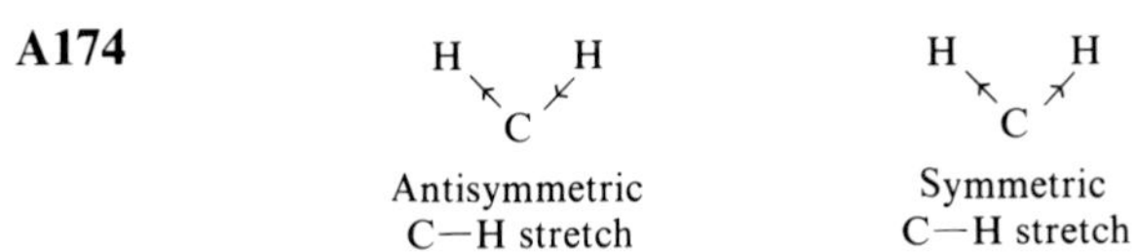

F176 The methyl group has three stretching vibrations. However, only two absorption bands are normally observed in spectra (**F69**).

The four diagrams below represent different stages of two vibrations. Arrange them in two pairs so that each pair describes one vibration.

(a) (b) (c) (d)

A175 (**3**). Both (**1**) and (**2**) have 'normal' methylene groups which would absorb near 2850 cm^{-1}.

F177 How many stretching bands would you expect to be associated with a methine hydrogen?

A176 (b) and (d) (symmetric); (a) and (c) (antisymmetric). In practice there are two antisymmetric vibrations but they cannot be distinguished and so the methyl group is commonly observed to have two CH stretching absorption bands (2962 and 2872 cm^{-1}).

F178 In summary, then, the 2800—3000 region in the spectrum of a saturated hydrocarbon will contain absorptions due to symmetric and antisymmetric $-CH_2$ stretches, symmetric and antisymmetric $-CH_3$ stretches and $\geqslant$C—H stretch. The performance of routine infrared spectrophotometers fitted with

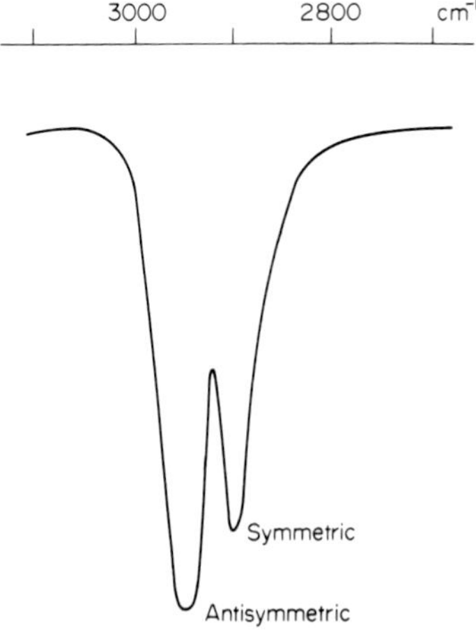

sodium chloride prisms is quite inadequate to resolve the CH_3 bands from the corresponding CH_2 bands and from the methine CH stretching absorption. The usual appearance of saturated CH stretching absorption is that of a doublet. Some modern grating instruments, however, have better resolution in this region.

Continue with the next frame.

A177 One. It occurs near 2890 cm^{-1}. It is very weak and generally of no diagnostic value.

F179 You will recall that in Part 1 it was stated that bending vibrations as a class are more complicated than stretching vibrations. We subdivided them into four distinct types. Write down their names.

F180 Of these, deformations of CH_3 and CH_2 have the highest frequency. Consider the following statements.

(i) Deformations occur virtually within the group.
(ii) Deformations would be expected to be very sensitive to the environment of the group.

Decide whether

(a) (i) is true and (ii) false;
(b) (i) is false and (ii) true;
(c) both are true;
(d) both are false.

A179 Deformation, wag, rock, twist.

F181 Most CH_2 and CH_3 rocking and wagging vibrations occur in the skeletal region of the spectrum. (Twists occur outside the range of most spectrophotometers.) Which type of vibration gives rise to the most reliable absorption bands below 1500 cm^{-1} in the spectra of hydrocarbons—deformations, rocks or wags?

A180 (a). Rocks and wags involve motion of the group as a whole relative to the rest of the molecule. These are much more affected by the environment. (See diagram in **F182**).

F182 The methyl group has two deformations, the symmetrical (~1380 cm^{-1}) and the antisymmetrical (~1450 cm^{-1}). Only one deformation is possible for the methylene group and this occurs near 1465 cm^{-1}.

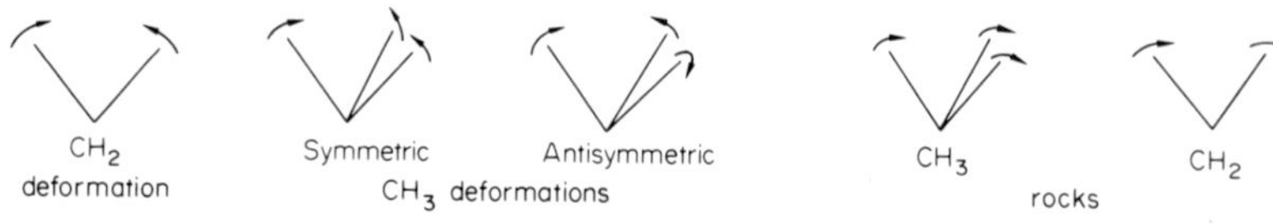

Part of the spectrum of *n*-pentane is given below. Describe the vibrations corresponding to each of the absorptions. A, B, and C.

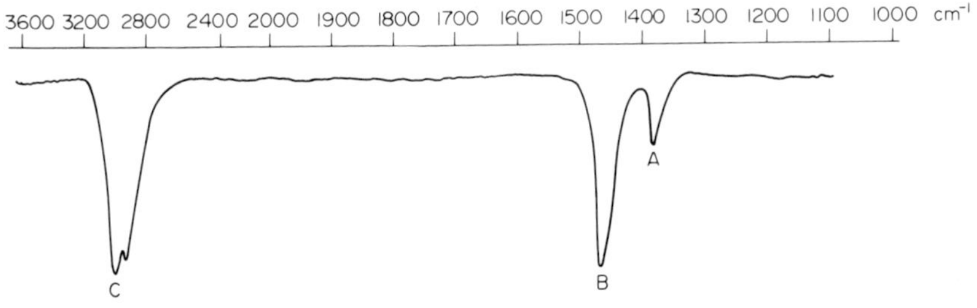

A181 deformations. Because of the rather indeterminate nature of rocking and wagging vibrations (which frequently involve C—C or C—O stretching as well) the term 'rock' (or 'wag') is often used rather loosely to describe bands which are thought to be associated with vibrations in which this type of motion makes an important contribution.

F183 Rocking and wagging vibrations give absorptions which are of little value except when four or more consecutive methylene groups occur in an open-chain. Compounds containing such a feature generally show a broad band near 720 cm^{-1} which is ascribed to the in-phase wagging of the $—CH_2—$ groups.

Which of the following compounds will show a band at 720 cm^{-1} in its spectrum?

4-methylheptane; *n*-pentane; 2-methyloctane; 2-methylhexane; 2,6-dimethyldecane.

A182 A: CH_3 symmetric deformation

B: CH_3 antisymmetric deformation
CH_2 deformation

C: CH_3 antisymmetric and symmetric CH stretch
CH_2 antisymmetric and symmetric CH stretch

F184 Another very useful correlation has been established for molecules containing the *gem*-dimethyl group $>C(CH_3)_2$. They all show a characteristic doublet near 1375 cm^{-1}. This is due to coupling between the symmetrical deformations of the two methyl groups. Assign each of the three compounds (**4**), (**5**) and (**6**) to one of the partial spectra (A), (B) and (C).

$CH_3—CH(CH_3)—CH_2—CH_3$

(**4**)

CH_2 ring: $CH_2—CH_2—CH_2—CH_2—CH_2—CH_2$ (cyclohexane)

(**5**)

$CH_3—CH_2—CH(CH_3)—CH(CH_3)_2$

(**6**)

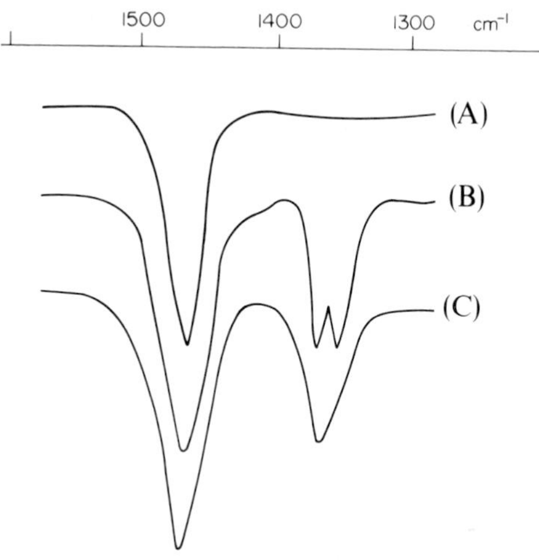

A183 2-methyloctane . None of the other compounds has more than three adjacent CH_2 groups: *n*-pentane ; 4-methylheptane ; 2-methylhexane ; 2,6-dimethyldecane . For this correlation to hold the band must lie very near 720 cm^{-1}. Bands may occur just above 725 cm^{-1} for compounds having three adjacent CH_2 groups e.g. *n*-pentane (726 + 730 cm^{-1}) and 3-methylheptane (728 cm^{-1}).

F185 Although saturated hydrocarbons are not a particularly important group of compounds, aliphatic C—H vibrations occur in practically all organic materials. The information you have been given and deduced for yourself so far in this section is therefore applicable to the majority of the spectra that you will meet. Moreover, the presence of a functional group does not generally affect the vibration of alkyl groups, and when it does only the CH bonds adjacent to the functional group are involved.

When examining the spectrum of an unknown material one can usually decide immediately whether or not aliphatic CH bonds are present provided the sample is pure. (The presence of trace impurities such as solvents or stop-cock grease will often produce bands near 2900 cm^{-1} even when the compound itself has no aliphatic CH groups.) Furthermore, after identifying the functional groups one may be able to establish the nature of the alkyl groups by a more detailed study of the CH bands.

A primary aliphatic alcohol is known to contain eight carbon atoms. Its spectrum contains, among others, absorption bands with following frequencies: 1367, 1382, 1460, 2865, 2940, 3050 cm^{-1}

There is no band at 720 cm^{-1}. Which of the following structures has the closest fit to these frequencies?

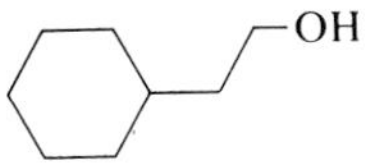

OH (7) OH (8) OH (9)

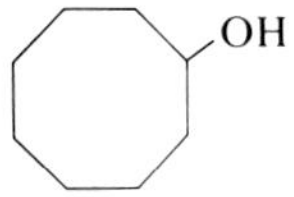

OH (10) OH (11) OH (12)

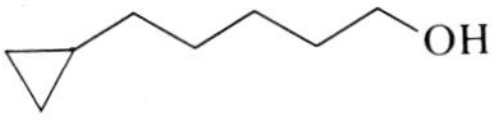

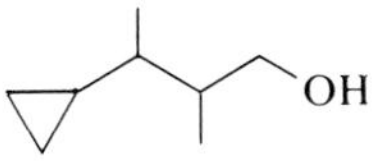

A184 (**4**) B; (**5**) A; (**6**) C. (See also **A182**.)

A185 (**10**). (**8**) and (**11**) are eliminated by the absence of a band at 720 cm^{-1}. The doublet 1367/1382 establishes the presence of *gem*-dimethyl groups favouring (**10**). Note also the band at 3050 cm^{-1} due to cyclopropyl C—H. This is a rather artificial problem. In practice one usually obtains information regarding the nature of alkyl groups from n.m.r. or mass spectra.

Unsaturated Hydrocarbons

F186 Write down the four types of vibration associated with the group $>C{=}C<^{H}$ in order of increasing frequency. (You should be able to do this without consulting Chart 2.)

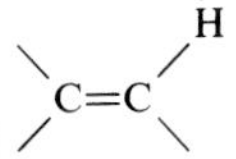

F187 Which type of vibration in the molecule H(R)C=C(R)H will not give rise to infrared absorption?

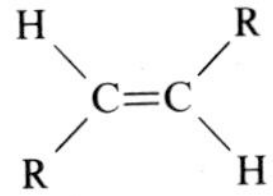

A186 out-of-plane C—H deformation < in-plane CH deformation < C=C stretch < C—H stretch.

F188 There are three ways in which two or more double bonds can occur in a molecule

(A) cumulated e.g. $(CH_3)_2C{=}C{=}C(CH_3)_2$

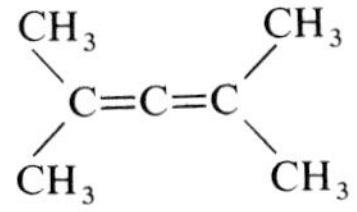

(B) conjugated e.g. $CH_3CH{=}CH{-}CH{=}CH_2$

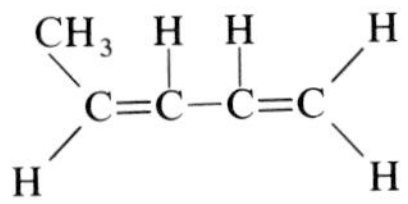

(C) separated e.g. $(CH_3)_2C{=}C(CH_3){-}(CH_2)_4{-}C(CH_3){=}C(CH_3)_2$

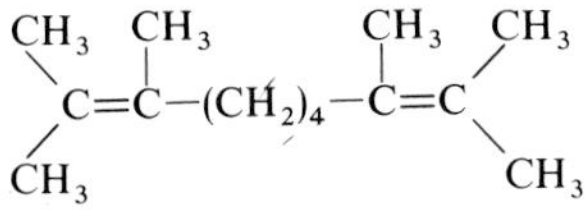

Assign each of the following descriptions of the 1500—2000 cm^{-1} region of the spectrum to the examples given for A, B or C:

(a) one absorption near 1650 cm^{-1};
(b) one absorption near 1950 cm^{-1};
(c) two absorptions near 1650 cm^{-1}.

A187 C=C stretch. In a symmetrical alkene there is no change of dipole moment during this vibration. Some CH vibrations will also be inactive due to symmetry but there will always be one vibration of each type which does absorb infrared radiation.

F189 Some useful empirical generalizations can be made about C=C stretching frequencies.

(i) Conjugation usually lowers the frequency. The intensity is also greatly enhanced when the C=C bond is conjugated with C=O.

(ii) Cyclohexenes have C=C stretching frequencies identical with their open chain analogues but as the ring size of an unsaturated cyclic compound gets smaller endocyclic double bonds decrease in stretching frequency whereas exocyclic double bonds increase in frequency. (See **A138** for explanation).

endocyclic   exocyclic

(iii) Chlorine, bromine and iodine tend to lower frequencies by conjugation of the double bond with the non-bonding electron pairs of the halogen atom. When fluorine is attached to a double bond, however, the latter has abnormally high frequency.

Arrange the following sets of double bonds in order of increasing C=C stretching frequency.

(a) 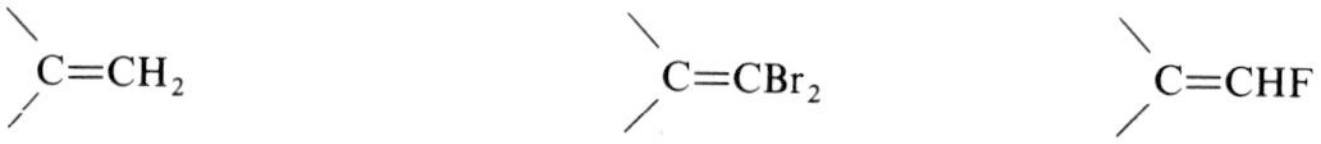 $>C=CH_2$ $>C=CBr_2$ $>C=CHF$

(b) 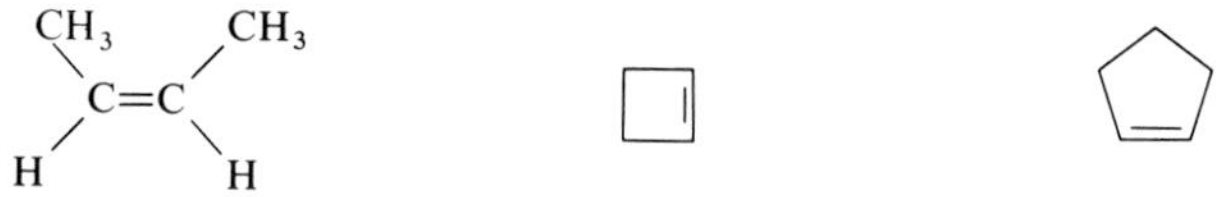 $CH_3CH=CHCH_3$ (cis)   cyclobutene   cyclopentene

(c) 2-methylenecyclohexanone   methylenecyclohexane   methylenecyclobutane

---

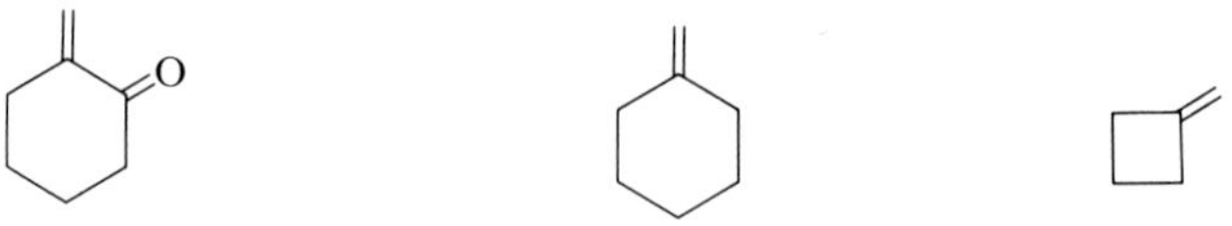

A188 (**C**)(a); (**A**)(b); (**B**)(c). (**C**) contains two identical but separated double bonds. The abnormally high frequency of cumulated double bonds (**A**)

is due to coupling and the vibration observed is the antisymmetric one. The symmetric stretch occurs at low frequencies and is usually of little value. (B) contains a monosubstituted and a disubstituted double bond.

F190 The in-plane CH bending vibrations occur near 1400 cm^{-1} and the out-of-plane bends are to be found in the region 650—1000 cm^{-1}. Which of the two types would you expect to be more useful in *recognizing the presence* of an alkenic group?

- - -

A189

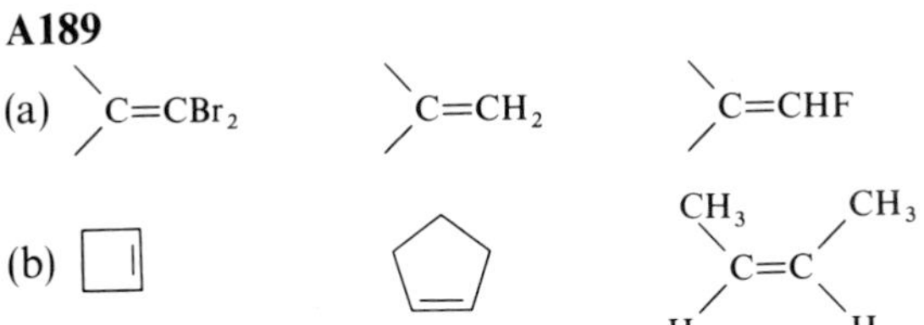

(*cis*-but-2-ene absorbs at the same frequency as cyclohexene)

(c) O

Incidentally, cyclopropenes are exceptional in that their C=C stretching frequencies are similar to that of cyclohexene.

F191 The out-of-plane vibrations of alkenes were extensively covered in Part 3. They are usually the strongest bands in the spectrum and yield the most information about the structure of the alkene. Which of the following pairs could **not** be distinguished by the out-of-plane CH deformation bands?

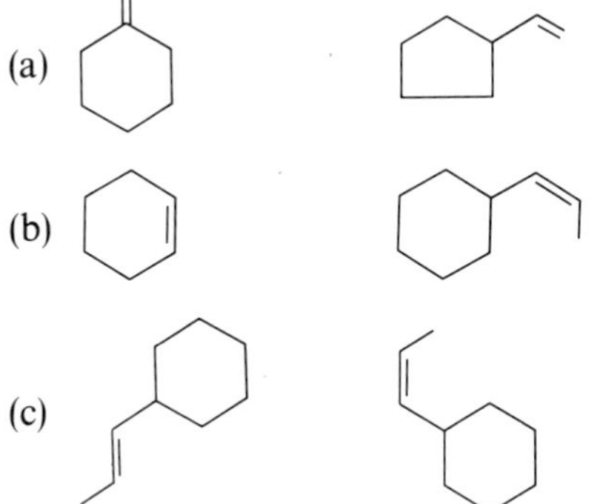

- - -

A190 out-of-plane. The in-plane bands could easily be masked by strong 'saturated' CH bending absorptions.

F192 Which of the following absorption patterns is most likely to be observed for cyclohexene?

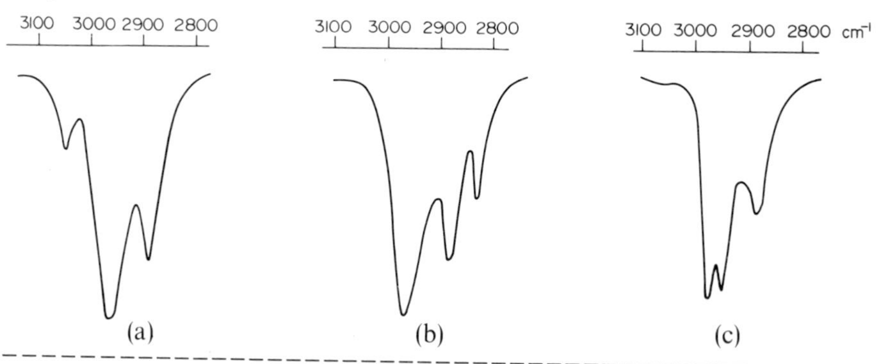

A191 (b). Both are *cis*-disubstituted.

F193 The alkynic CH stretch (≡C—H) occurs at even higher frequency (3300 cm^{-1}) and like other CH stretching absorptions it appears as a narrow, sharp band. The frequency lies within the range found for hydrogen-bonded O—H and N—H (later).

Under which of the following circumstances could you categorically say that a ≡C—H bond is present in a molecule? The infrared spectrum shows

(a) no absorption above 3100 cm^{-1};
(b) a broad absorption near 3300 cm^{-1};
(c) a sharp absorption at 3600 cm^{-1} but no other above 3100 cm^{-1};
(d) sharp absorptions at 3300 cm^{-1} and 3600 cm^{-1}.

A192 (a). =C—H stretching is higher frequency than ⩾C—H (3000—3100 cm^{-1}).

F194 The —C≡C— stretching absorption of alkynes occurs at 2150 cm^{-1}. However, it is generally very weak. Which of the following alkynes would you expect to have the strongest C≡C stretching absorption?

CH_3—C≡C—CH_2CH_3	CH_3—C≡C—CH_3	CH_3—C≡C—H
(13)	**(14)**	**(15)**

A193 (d). In (b) the ≡CH absorption might be hidden beneath the broad band.

F195 ≡CH bending of acetylene gives rise to a strong broad absorption in the range *1900—1800/1600—1500/700—600 cm*$^{-1}$. Which of the compounds **(13)**, **(14)** and **(15)** in **F194** would not show this feature?

A194 (15). $CH_3—C\equiv C—H$ would have the most polar $C\equiv C$ bond. $C\equiv C$ would be inactive in (**14**).

F196 The inclusion of a benzenoid ring in an organic compound will give rise to infrared absorption in the following regions.

(i) 3000—3100; C—H stretch; weak–moderate (intensity)
(ii) 1650—2000; combination and overtone bands; weak
(iii) 1550—1600; } ring stretching vibrations { weak–moderate
(iv) 1450—1500; } ring stretching vibrations { moderate
(v) 1000—1300; C—H in-plane bend; weak
(vi) 600—900; C—H out-of-plane bend; strong

Analogous vibrations in alkenes give rise to absorptions in regions (i), (iii) and (vi) whilst the in-plane CH bending vibration (v) absorbs within the fingerprint region and is consequently of little value. Although alkenes can give rise to weak absorption in region (ii), the feature is by no means as prominent as in simple benzenoid compounds. In complex benzenoid compounds, however, the absorption becomes very diffuse and difficult to distinguish from an uneven base-line. This leaves us with region (iv). In practice, a moderate to strong absorption near 1500 cm^{-1} (in addition to absorption in regions (i), (iii) and (vi)) is perhaps the most reliable feature in establishing the presence of a benzenoid ring, although, if it lies at the low frequency end of the region it could be obscured by absorption due to methylene and methyl bending vibrations.*

Select from Fig. 4.1 the two spectra which contain evidence for the presence of a benzenoid ring and for each write down the frequencies for =CH stretching, ring stretching and =CH out-of-plane bending vibrations.

* Benzenoid compounds have four bands within regions (iii) and (iv) (near 1450, 1500, 1570 and 1600 cm^{-1}) although not all may be seen in any one compound.

A195 600—700 cm^{-1}. (**13**) and (**14**).

F197 Once the presence of a benzenoid ring has been established, which of the regions (i) to (vi) would you examine further in seeking information on the number and relative positions of the substituents?

A196 (b) and (c).

	=CH stretching	C=C stretching	=CH out-of-plane bending
(b)	3100	1580 and 1600	700
(c)	3050	1520 and 1600	820

The doublet at 1580/1600 shown in spectrum (b) is highly characteristic. The 1500 cm^{-1} band is missing in this spectrum but the 1450 cm^{-1} band taken in conjunction with those near 3100 cm^{-1} and 700 cm^{-1} also indicate a benzenoid ring.

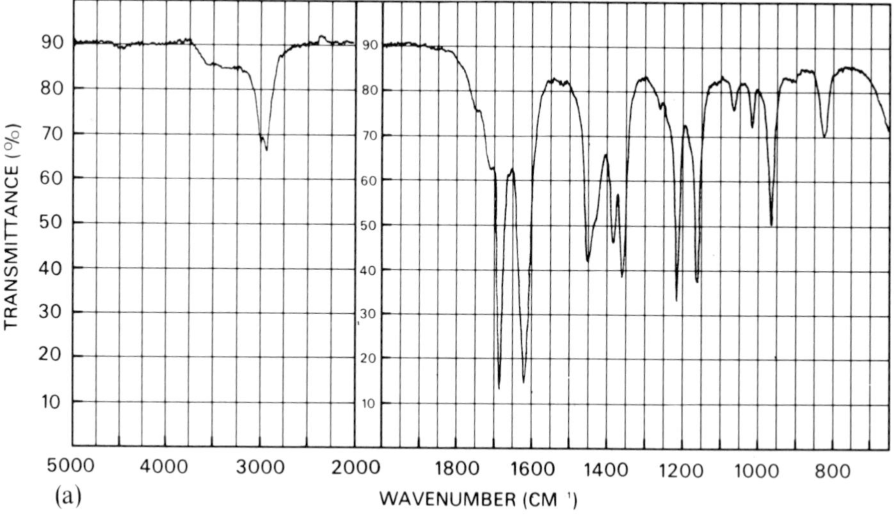

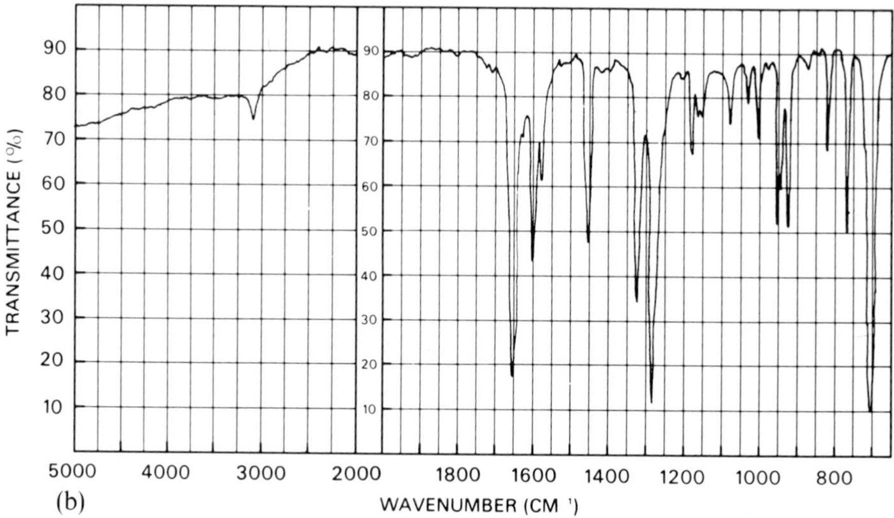

Fig. 4.1. (F196).

In (a) the band near 1600 cm^{-1} is at too high a frequency and too strong for a benzenoid absorption. The absence of the 1500 cm^{-1} band in conjunction with the absence of absorption in the 650 to 900 cm^{-1} region also suggests the absence of a benzenoid ring. Similar arguments show (c) to be benzenoid and (d) to be aliphatic.

A197 (vi). This was dealt with in frames 118—124 of Part 3.

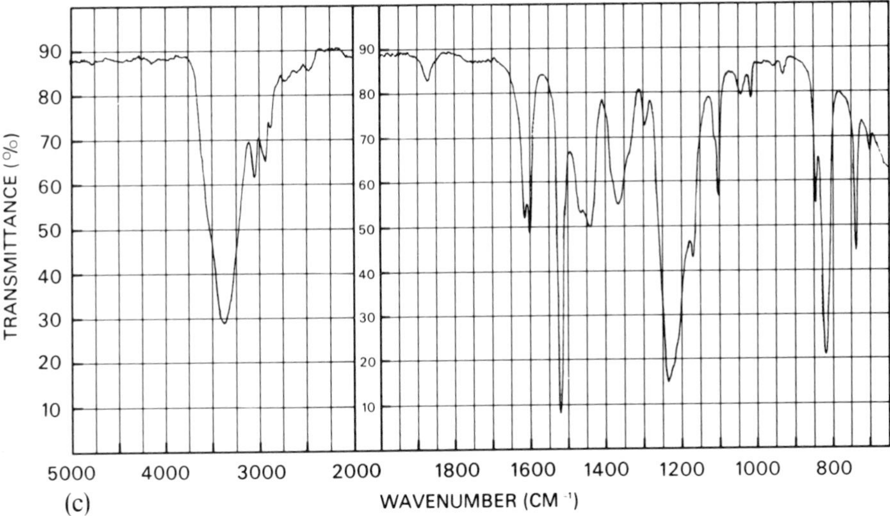

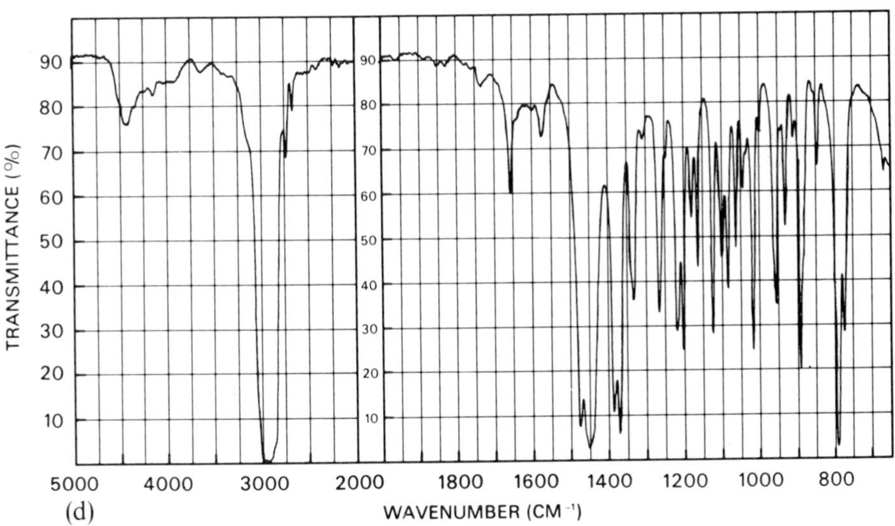

Fig. 4.1—*contd.*

Alcohols, Phenols, Amines and Ethers

F198 Alcohols and phenols, R—OH and Ar—OH, have three vibrations pertaining to the hydroxyl function.

(i) O—H stretch;
(ii) C—O stretch;
(iii) C—O—H bend.

The bending mode (iii) is usually associated with skeletal vibrations and absorbs in the 'fingerprint' region. Consequently it does not produce a band which may be considered diagnostic of hydroxyl. Although the C—O stretch is also coupled to other vibrations and absorbs in the fingerprint region, bands involving changes in C—O bond length can usually be identified. Which of the following alternatives makes the identification possible?

(a) C—O stretching bands invariably appear as doublets;
(b) C—O stretching bands are usually very strong;
(c) unlike the bending modes, C—O stretching frequencies cover a very narrow range.

F199 The free O—H stretching vibration has one of the highest frequencies encountered in organic compounds ($3650 \pm 50\,\mathrm{cm}^{-1}$). However, in solution and in the solid state, hydrogen-bonding between molecules occurs to a variable extent within the sample so absorption occurs over a range of frequencies giving rise to broad bands.

The following spectra were recorded for a compound at different concentrations in carbon tetrachloride. Which one corresponds to the most concentrated solution?

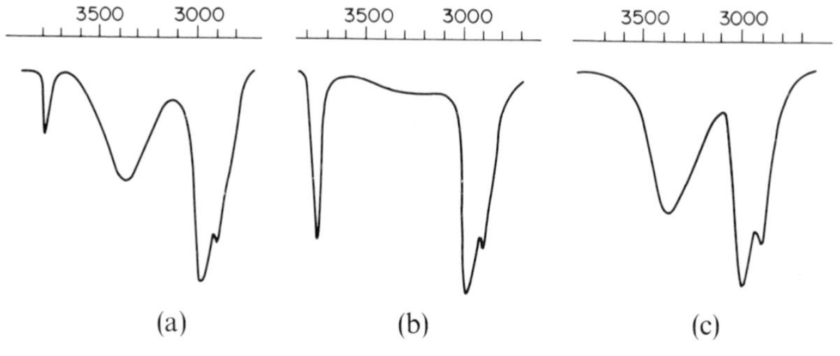

(a) (b) (c)

- -

A198 (b). (**F115** and **F116**).

F200 The spectra of the two phenols (**16**) and (**17**) are given at the top of p. 123.

(**16**) (**17**)

Each compound was measured as a solution in carbon tetrachloride and the molar concentrations were identical. Which spectrum corresponds to (**17**)?

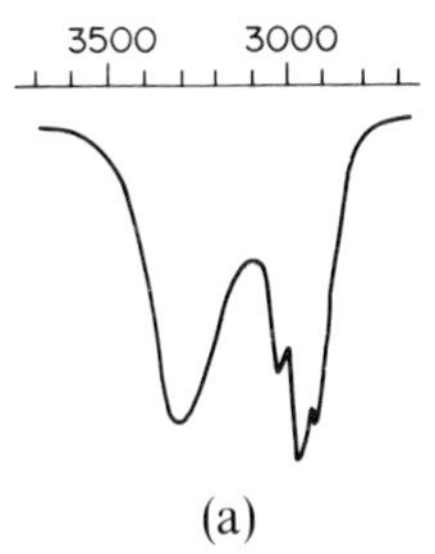

(a)

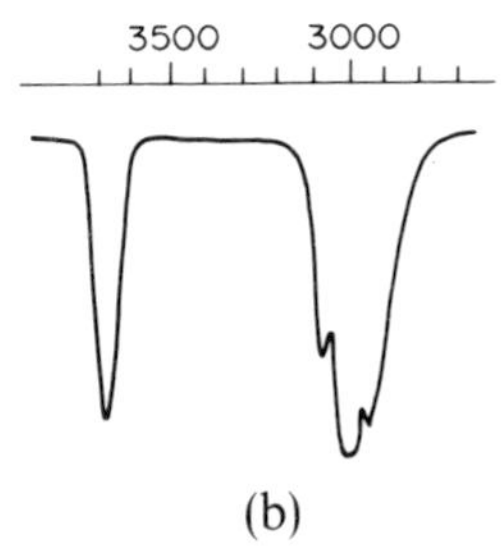

(b)

A199 (c). Hydrogen bonding between molecules becomes more important as the concentration increases and accounts for the lower frequency band. Spectrum (b) corresponds to the most dilute solution.

F201 The position of the OH band can also be used to distinguish between intermolecular and intramolecular hydrogen bonding, for example in

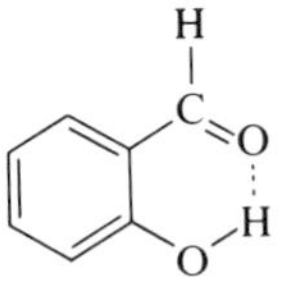

Which of the following spectra provides the best evidence of intramolecular hydrogen bonding?

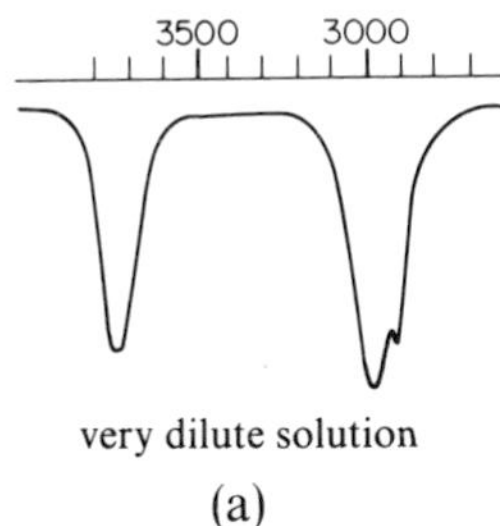

very dilute solution
(a)

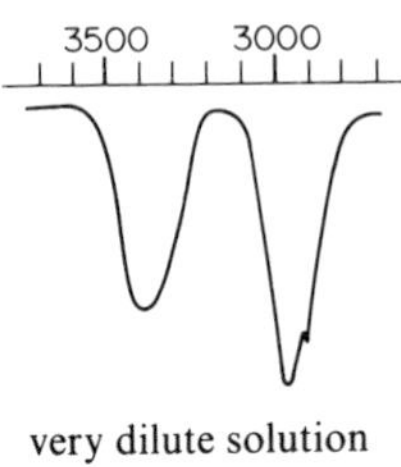

very dilute solution
(b)

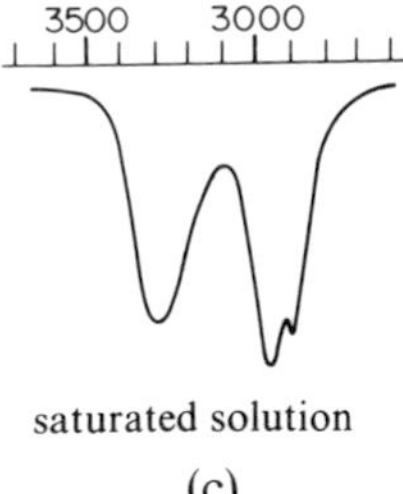

saturated solution
(c)

A200 (a). The large bulky groups in (**16**) prevent the hydroxyl hydrogen from approaching the hydroxyl oxygen of another molecule, hence it behaves as a 'free' OH group.

F202 The N—H stretching vibrations of primary and secondary amines are similarly affected by hydrogen bonding, although not to the same extent. Thus the highest frequency for N—H stretching of a secondary amine will be observed when the compound is measured in *very dilute/saturated* solution.

A201 (b). Absorption at a relatively low frequency in **dilute** solution indicates that hydrogen-bonding is independent of concentration. This is characteristic of intramolecular H–bonding.

F203 How would you distinguish between the N—H stretching absorption of a primary amine (**A**) and a secondary amine (**B**)?

A202 very dilute.

F204 Free and hydrogen-bonded N—H stretching frequencies occur within the region of hydrogen-bonded O—H. The former can usually be distinguished, however, by their relative sharpness, even when hydrogen bonded.

Which of the following spectra was obtained from an amine?

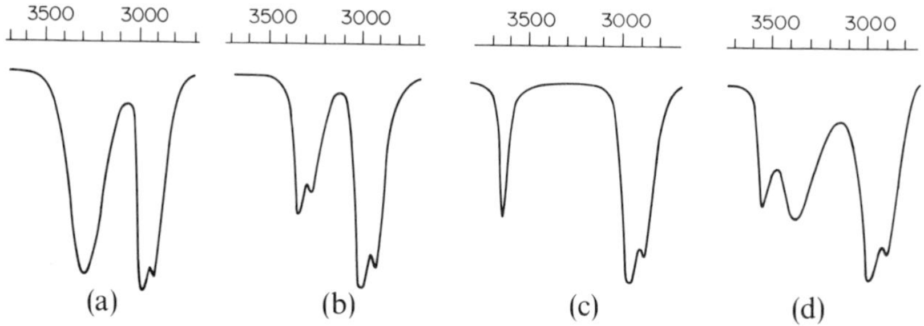

A203 N—H absorption of a primary amine would appear as a doublet, the two components corresponding to symmetric and antisymmetric modes. The separation is usually about 100 cm^{-1} and would be independent of concentration. N—H absorption of a secondary amine would generally have a single maximum although hydrogen-bonding could give rise to two, giving the band a similar appearance to that of a primary amine. However, the hydrogen-bonded component would disappear as the concentration of the solution was reduced. Two bands might also be observed for molecules having more than one NH (or OH) group.

F205 The N—H bending vibration of primary amines normally has a frequency in the range of 1580—1650 cm^{-1}. A selection of the bands in the two spectra shown in Fig. 4.2 have been labelled A to H. Which of these correspond to:

(a) N—H bend; (b) C—O stretch?

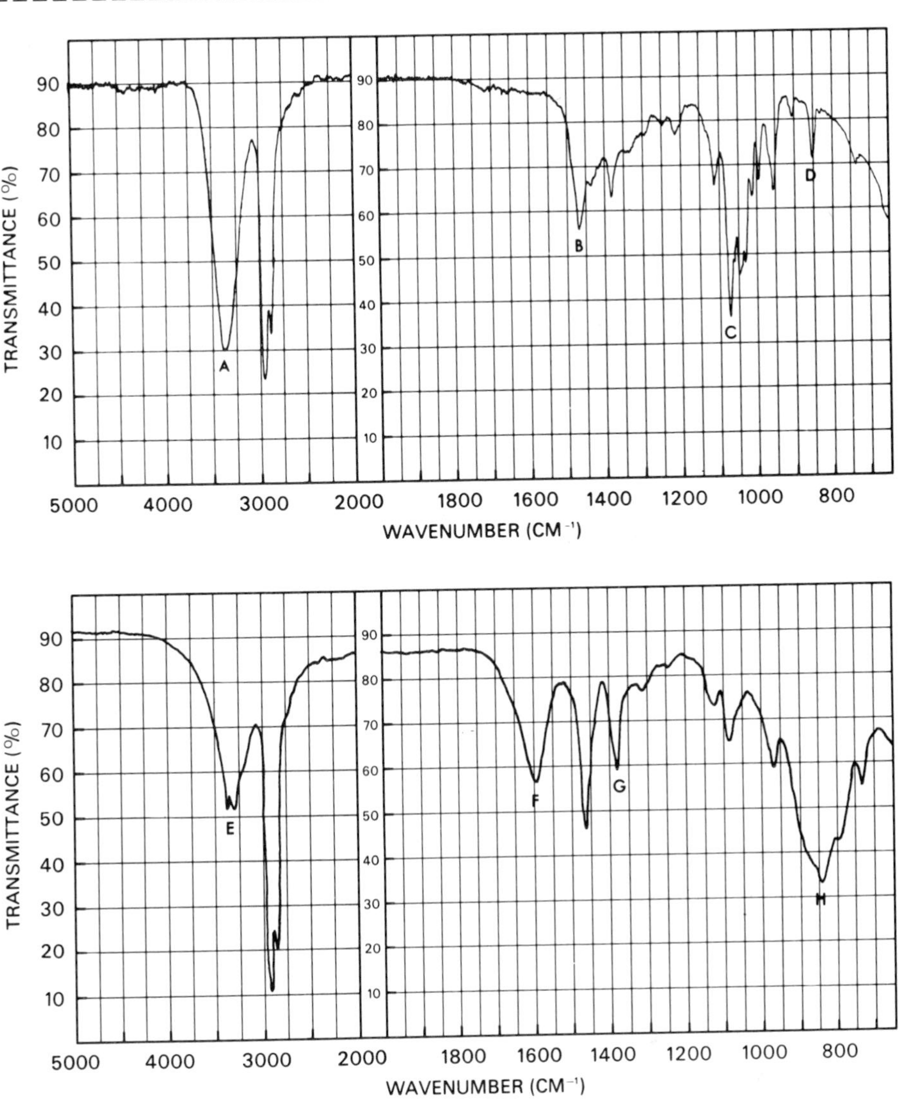

Fig. 4.2

A204 (b). The absorptions in (a) and (d) are too broad to be N—H bands. (c) is outside the range for N—H (it is characteristic of 'free' OH). (d) is typical of a hydroxyl compound which contains some isolated molecules and some hydrogen bonded ones.

F206 Ethers, R—O—R, show strong absorption in the fingerprint region but, because it is due to skeletal vibrations involving the C—O bond, the frequency is variable (1000—1250 cm^{-1}). Summarize how you would use infrared spectroscopy to determine whether the oxygen atom in the compound $C_6H_{10}O$ was present as a carbonyl, hydroxyl or ether linkage.

A205 Examination of the spectra above 3000 cm^{-1} indicates the upper one contains OH and the lower NH_2. Hence C is the C—O stretching absorption and F the NH bending absorption.

A206 Carbonyl would give a strong absorption between 1650 and 1800 cm^{-1}. Hydroxyl would give a strong absorption between 3100 and 3650 cm^{-1} **together with** strong absorption between 1000 and 1250 cm^{-1}. Ethers would give strong absorption between 1000 and 1250 cm^{-1}.

Ketones and Aldehydes

F207 Which is the most informative feature in the spectrum of a ketone?

(a) intensity of the C=O stretching band;
(b) frequency of the C=O stretching band;
(c) intensity of bands associated with stretching vibration of the C—CO bond;
(d) frequency of the deformation vibration involving the

C
 \
  CO angle.
 /
C

F208 Try to recall (or estimate) the frequency range over which the ketonic stretching vibration may be found.

A207 (b). See Part 3, **F125—F141**.

F209 An open-chain, saturated aliphatic ketone will absorb at 1715 cm^{-1}. However there are a number of structural effects which will lead to values well above and below this figure. We discussed them at length in Part 3. For example, conjugation with an alkenic double bond or aromatic ring results in a *decrease/increase* of a ketone carbonyl stretching frequency. A decrease in the angle θ,

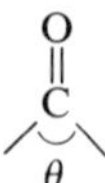

leads to an *increase/decrease*.

A208 1620—1800 cm^{-1}.

F210 Place the following compounds in order of increasing carbonyl stretching frequency.

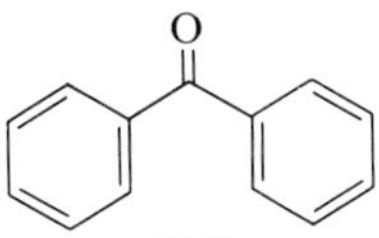

(18) 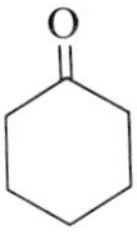 (19) 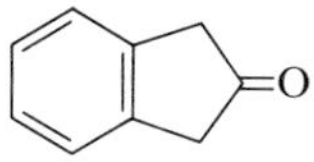 (20)

A209 decrease (**F136**), increase (**F138**).

F211 Hydrogen bonding can also influence the carbonyl frequency of a ketone, particularly when geometry allows intramolecular interaction in, for example, hydroxy- and amino-ketones.

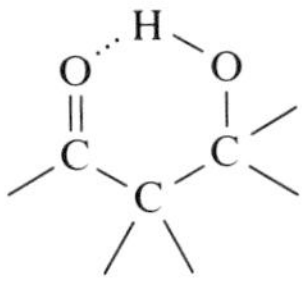

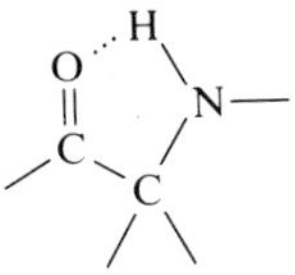

Suggest in one or two sentences an experiment involving infrared spectroscopy which would enable you to recognize this feature in a molecule.

A210 The order given is correct:

(**18**) conjugated;
(**19**) six-membered ring ketone, hence same frequency as open chain;
(**20**) five-membered ring ketone, not conjugated.

F212 Unlike monoketones, β-diketones frequently exist as mixtures of enolic and ketonic forms.

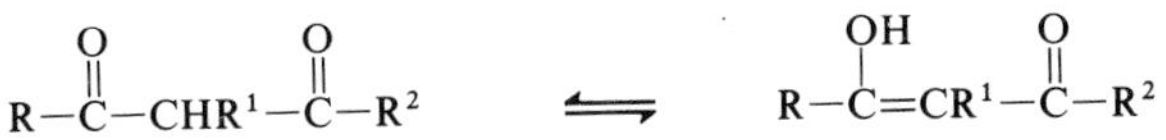

$$\underset{\text{ketonic}}{R{-}\overset{\overset{\displaystyle O}{\|}}{C}{-}CHR^1{-}\overset{\overset{\displaystyle O}{\|}}{C}{-}R^2} \rightleftharpoons \underset{\text{enolic}}{R{-}\overset{\overset{\displaystyle OH}{|}}{C}{=}CR^1{-}\overset{\overset{\displaystyle O}{\|}}{C}{-}R^2}$$

This is because the enolic form is stabilized by hydrogen bonding;

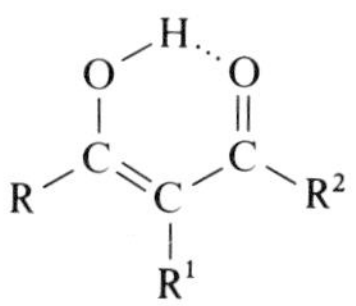

The spectrum in Fig. 4.3 is that of acetylacetone, $CH_3COCH_2COCH_3$. Deduce which vibrations will give rise to absorptions A, B and C.

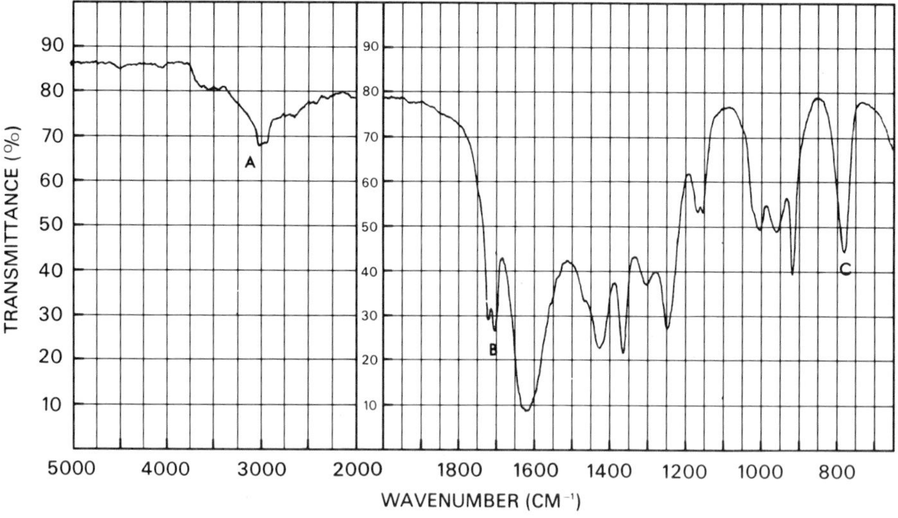

Fig. 4.3. Spectrum of acetylacetone.

A211 Run the infrared spectra of solutions of the compound at different concentrations. If absorption in the hydrogen bonded region of the O—H and N—H stretching vibration is unaffected by concentration then intramolecular hydrogen-bonding is responsible.

F213 Is the following statement true or false? The stretching frequency of a ketonic carbonyl group is dependent on a number of factors, among which are ring strain, conjugation and hydrogen bonding.

A212 A, C—H and O—H stretch;
B, C=O stretch of the keto-form;
C, =C—H out-of-plane bend of the enol form.

F214 Ketones always show absorptions of medium intensity between 1100 and 1300 cm^{-1}. Which of the following types of vibration most accurately describes the origin of these bands?

(a) Vibrations involving C—C stretching and CH bending (skeletal);
(b) C—C stretching vibrations;
(c) C—C—C bending vibrations.

A213 True.

F215 Record likely frequencies for such skeletal vibrations in acetylacetone (Fig. 4.3).

A214 (a). See Part 3 (**F116**). The fact that these bands are of medium intensity enables them to be distinguished from the much stronger absorption in the same region due to C—O (later).

F216 Methyl groups are electron donating relative to hydrogen atoms. As a consequence, the carbonyl stretching vibration of an aldehyde has a slightly *higher/lower* frequency than that of a corresponding methyl ketone.

A215 1160 (doublet) and 1250 cm^{-1}.

F217 Assign each of the following frequencies to one of the structures given below.

1730 1768 1700 1666 cm^{-1}.

C_6H_5CHO	CH_3CHO	$C_6H_4(OH)CHO$	Cl_3CCHO
(**21**)	(**22**)	(**23**)	(**24**)

A216 higher. Electron donation to the carbonyl group leads to more single bond character and hence a lower stretching frequency in ketones.

F218 The most characteristic feature of an aldehyde spectrum which enables distinction from a ketone is the former's unique C—H stretching frequency. Which of the two spectra, Fig. 4.4 or Fig. 4.5, corresponds to butyraldehyde

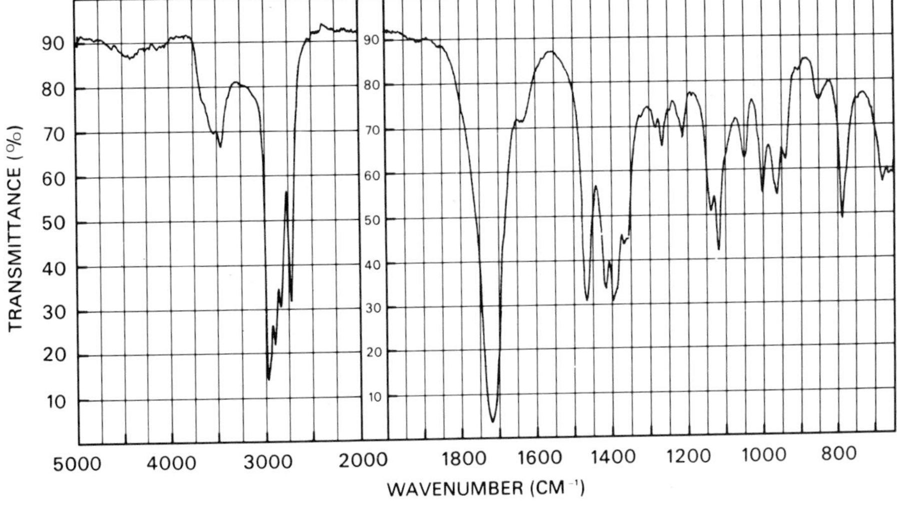

Fig. 4.4

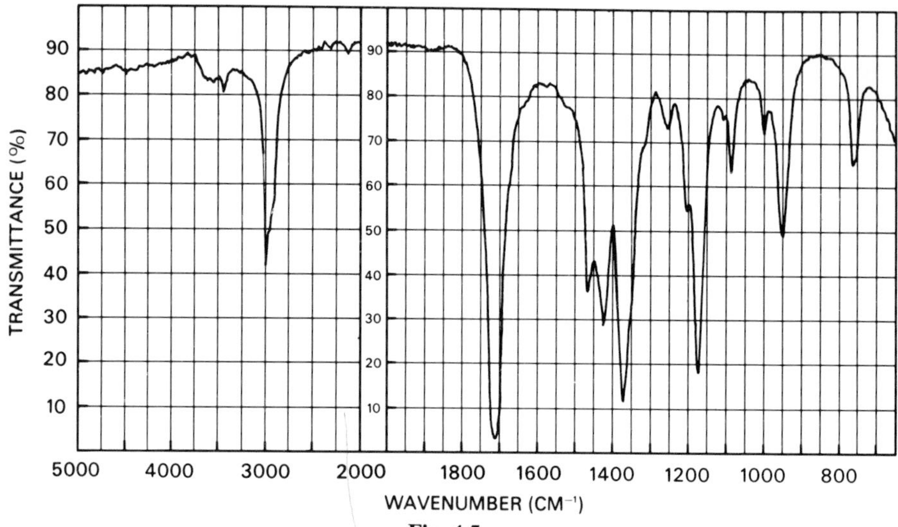

Fig. 4.5

$CH_3(CH_2)_2CHO$? The other one is 2-butanone, $CH_3COCH_2CH_3$. Record the aldehyde CH stretching frequency.

A217 (**21**), 1700; (**22**), 1730; (**23**), 1666; (**24**), 1768 cm^{-1}. Conjugation lowers frequency as does hydrogen bonding (**23**). The three relatively electronegative chlorine atoms tend to make the carbonyl group shorter and less polar in (**24**) than in (**22**).

F219 The aldehydic CH absorption band frequently appears as a doublet. (In Fig. 4.4 the second component is clearly visible on the side of the CH band at about 2830 cm^{-1}). This is attributed to Fermi resonance between the fundamental CH stretching mode and the first overtone of another vibration. Write down the fundamental frequency of the latter in benzaldehyde (Fig. 4.6) and use this, together

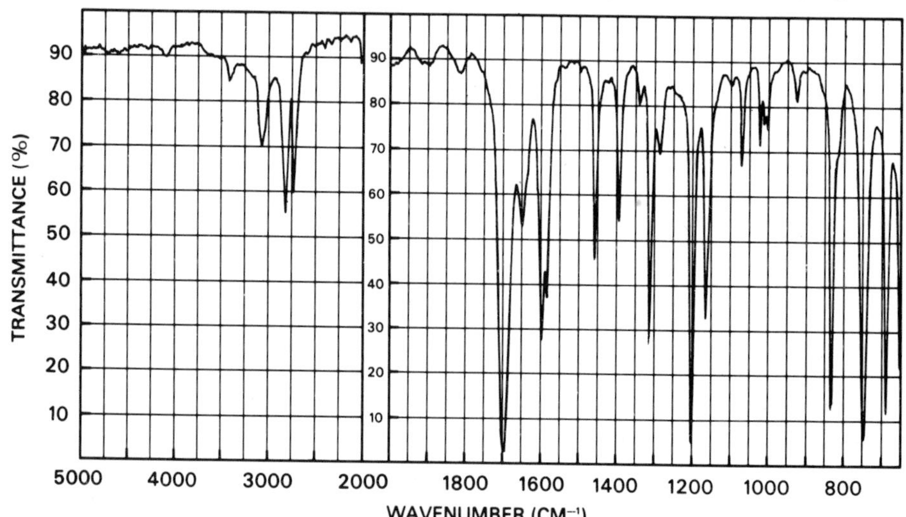

Fig. 4.6. Spectrum of benzaldehyde.

with your knowledge of conditions of Fermi resonance, to describe the vibration responsible.

A218 Fig. 4.4; 2750 cm^{-1}.

A219 1390 cm^{-1}. C—C(=O)H, C—C—H in-plane bend. 2 × 1390 = 2780 cm^{-1}. Both vibrations take place in the plane of the C(=O)H group. (see **F78**).

Carboxylic Acids

F220 When a carbonyl group is bound to an oxygen atom, two opposing effects are present, both of which strongly influence carbonyl frequency.

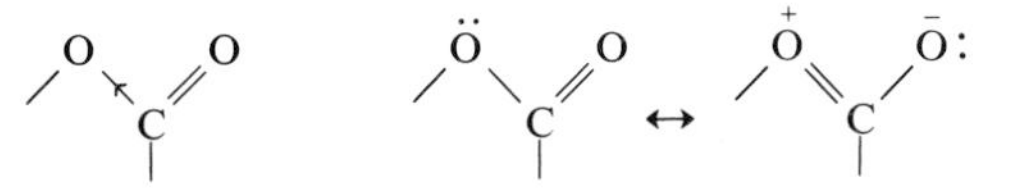

Negative inductive effect — Increases frequency

Resonance effect — Decreases frequency

Monomeric, saturated, aliphatic carboxylic acids absorb near 1760 cm^{-1}. By comparison with saturated aliphatic ketones, deduce which effect is dominant in carboxylic acids.

F221 Under normal conditions, however, such compounds are found to absorb near 1710 cm^{-1}. This is because in the solid state, and to a large extent in solution, carboxylic acids exist as hydrogen-bonded dimers. Each molecule is mutually hydrogen-bonded to another resulting in an overall symmetrical structure. See if you can draw it for the general case

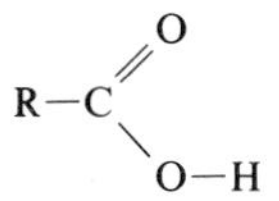

R—C(=O)—O—H

A220 The inductive effect. Electron withdrawal from the carbonyl group increases its bond order and leads to higher frequency. The resonance effect implies a significant contribution from a form in which the carbonyl group has a single-bond character. This would lead to a lengthening of the bond and consequently lower frequency than the corresponding ketone.

F222 Explain briefly in your own words the grounds for the assertion that such a dimer would have a lower C=O stretching frequency than an 'isolated' carboxylic acid molecule.

A221

```
       O···H—O
      //       \
  R—C            C—R
      \        //
       O—H···O
```

F223 The conjugation of multiple bonds with carboxylic acid dimer carbonyl groups leads to small shifts to lower frequency. The effect is much less noticeable than in analogous aldehydes and ketones. Try to deduce which of the following ranges is appropriate to the carbonyl absorption for the dimer of aryl conjugated acids without reference to the charts.

(a) 1710—1720 cm^{-1};
(b) 1680—1715 cm^{-1};
(c) 1650—1700 cm^{-1}.

A222 Hydrogen bonding to the oxygen atom of a carbonyl group leads to greater polarization of the group. This gives the bond more single bond character and reduces the frequency.

F224 The most diagnostic feature of a carboxylic acid spectrum is a very broad absorption, frequently extending from 2500 cm^{-1} to 3300 cm^{-1}. Which vibration is responsible?

A223 (b). This represents a shift to lower frequency which is somewhat less than one would expect for conjugated ketones (1660—1700 cm^{-1}).

F225 Which of the following factors is responsible for the diffuse shape and relatively low frequency?

(a) inductive withdrawal of electrons by the carbonyl group;
(b) Fermi resonance analogous to that observed with aldehydes;
(c) hydrogen bonding.

A224 O—H stretching. Similar bands are observed with amine salts, $RNH_3^+X^-$, but these will rarely possess the accompanying carbonyl band, present in the carboxylic acid spectrum.

F226 Partial spectra of benzoic acid and pentanoic acid ($CH_3(CH_2)_3 \cdot CO_2H$) are given below. Which is which?

A225 (c). Hydrogen bonding leads to a range of O—H bonds in the sample each in a slightly different environment and having a slightly different frequency. The observed band is a superposition of all the absorptions.

F227 The following three vibrations can also give rise to recognizable absorptions in the spectra of carboxylic acids: C—O stretch; C—O—H in-plane bend; C—O—H out-of-plane bend. Which of these are most likely to couple with the stretching of the C—COOH bond?

A226 (b), benzoic acid; (a), pentanoic acid. The latter shows aliphatic C—H stretching superimposed on the O—H band. Aromatic C—H stretching is much less intense and is hardly discernible here.

F228 The frequencies of the two bending vibrations are similar to analogous alkenic C—H modes. Assign each of the labelled absorptions in Fig. 4.7 (pentanoic acid) to one of the following vibrations. Mainly C—O stretch, C=O stretch, O—H stretch, mainly OH in-plane bend, OH out-of-plane bend.

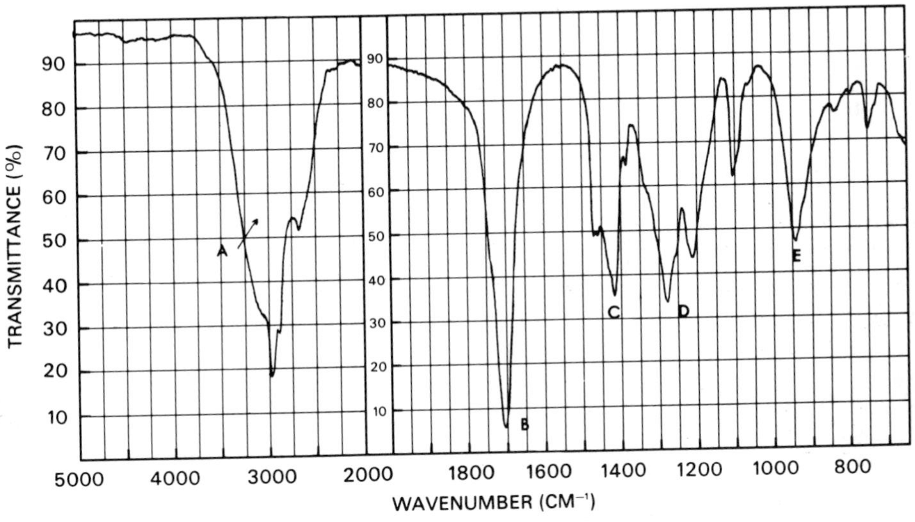

Fig. 4.7. Spectrum of pentanoic acid.

A227 The first two. Recall that one of the conditions for the coupling of vibrations is that both should have some motion in a common direction.

F229 The carbonyl absorption of the carboxylate anion differs considerably from that of the parent acid.

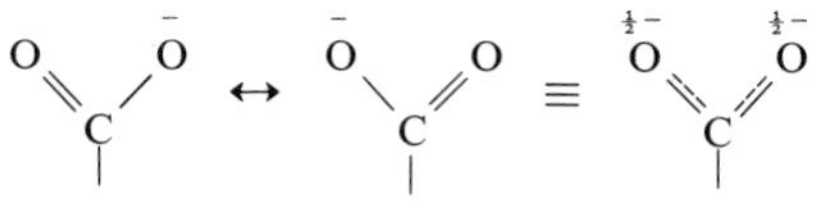

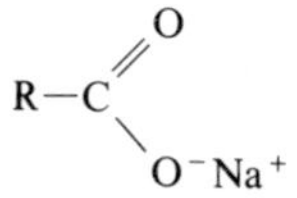

The group is symmetrical due to resonance, and the two CO bonds are of equivalent strength, intermediate between C=O and C—O. Which of the following descriptions of the absorption characteristics of the carboxylate anion would you expect to be correct?

(a) very intense single absorption at 1500 cm^{-1};
(b) two bands, one near 1600 and the other near 1400 cm^{-1};
(c) a very intense single absorption at 1800 cm^{-1}.

A228 A, O—H stretch
B, C=O stretch
C, mainly O—H in-plane bend*†
D, mainly C—O stretch†
E, OH out-of-plane bend*

* Compare the analogous vibrations in alkenes (acid dimer structure is planar).
† Participates in skeletal vibrations.

A229 (b). The situation is analogous to methylene $—CH_2—$ which has antisymmetric and symmetric C—H stretching. Thus we have

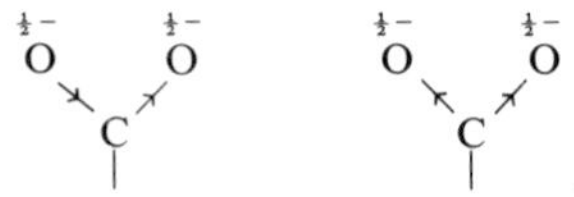

antisymmetric symmetric

Carboxylic Esters and Lactones

F230 Carboxylic esters show two strong absorptions. Which vibrations are responsible?

F231 The carbonyl absorption of some esters occurs within the ketonic and carboxylic acid carbonyl regions. In such cases, how would you distinguish an ester from a ketone or a carboxylic acid?

A230 C=O stretching; C—O stretching. Both C=O and C—O bonds are much more polar than any other bonds in the molecule.

F232 The carbonyl group of saturated esters absorbs in the 1735—1750 cm^{-1} region. Unsaturated esters with conjugated carbonyl groups absorb between *1715 and 1730* cm^{-1}*/1750 and 1760* cm^{-1}.

A231 An ester would show a strong absorption due to C—O. A ketone would absorb in this region due to skeletal vibrations but these bands would be of moderate intensity only (**F214**), and they are usually easily distinguished from the much stronger C—O band. Of course, ambiguity could arise in the case of a ketone which also possessed a C—O bond. An ester would not show the broad O—H stretching band of carboxylic acids.

F233 However, if the unsaturated moiety in the molecule is directly attached to the single bonded oxygen of the ester, then an **increase** in carbonyl frequency is observed. One possible explanation for this (which might help you remember the empirical observation) is that the oxygen lone pair electrons are involved in resonance with the carbon–carbon multiple bond in addition to the C=O group. Since resonance with the C=O group leads to a lowering of frequency, loss of this resonance through sharing with another multiple bond would lead to an increase in C=O stretching frequency.

Cyclic esters are known as lactones, and those with six-membered rings (δ-lactones) absorb at the same frequency as analogous open-chain compounds. Assign each of the frequencies 1720, 1745 and 1760 cm^{-1} to each of the following δ-lactones.

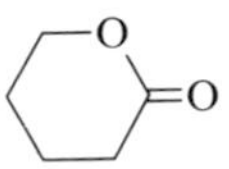

(25) 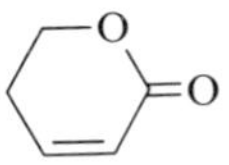 **(26)** 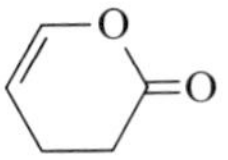 **(27)**

A232 1715–1730 cm^{-1}. Conjugation lowers the carbonyl frequency.

F234 Place the following lactones in order of **decreasing** carbonyl frequency.

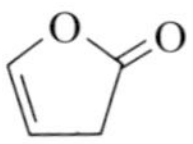

(28) 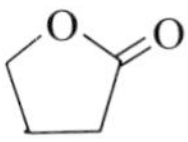 **(29)** 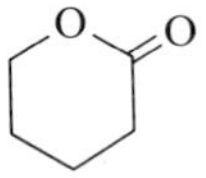 **(30)**

A232 **(25)** 1745; **(26)** 1720; **(27)** 1760 cm^{-1}. Re-read the frame carefully if you obtained the wrong answer.

F235 The frequency of the C—O stretching vibrations in esters and lactones varies considerably with structure because of coupling with C—C stretching and CH bending vibrations. Is the following statement true or false?

'C—O vibrations are skeletal and as such their frequencies are much less reliable in characterization than C=O vibrations'.

A234 **(28)**, **(29)**, **(30)**. A double bond adjacent to the singly bound oxygen atom of an ester or lactone raises the C=O stretching frequency. Five-membered ring carbonyl compounds absorb at higher frequency than six-membered ring compounds.

F236 However, C—O absorptions can always be recognized because of their
(a) sharpness,
(b) diffuseness,
(c) high intensity.

- -

A235 True.

A236 (c). See **A230**.

Acid Halides, Anhydrides and Amides

F237 Which of these three classes of carbonyl compounds has the highest carbonyl stretching frequency?

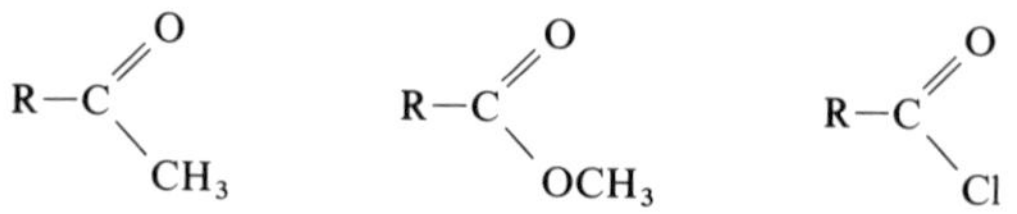

$R{-}C({=}O){-}CH_3$ $\quad$ $R{-}C({=}O){-}OCH_3$ $\quad$ $R{-}C({=}O){-}Cl$

- -

F238 In fact, saturated acid chlorides are among those compounds with the highest carbonyl frequency (1780—1815 cm^{-1}). At which of the following frequencies would you expect acid fluorides to absorb?
(a) 1860 cm^{-1}; (b) 1760 cm^{-1}.

- -

A237 $R{-}C({=}O){-}Cl$ This was discussed fully in Part 3, **F131—F135**.

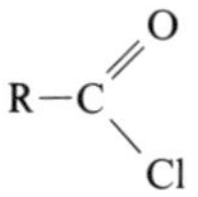

F239 Carboxylic anhydrides also have carbonyl bands at high frequencies but are easily distinguished from most other carbonyl compounds in that the absorption is almost invariably a doublet corresponding to symmetrical and antisymmetrical stretching modes.

(A) 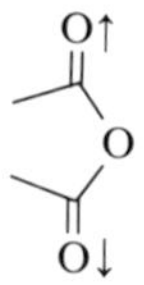 O↑ … O↓ $\quad$ (B) 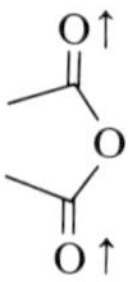 O↑ … O↑

Which is the antisymmetrical mode?

- -

A238 (a). Fluorine is more electronegative than chlorine.

F240 Assign each of the following pairs of absorptions to one of the compounds below. 1865, 1780; 1815, 1750; 1775, 1720.

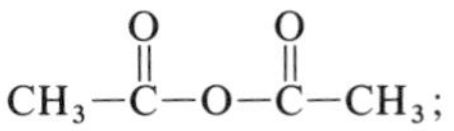

$CH_3-\overset{O}{\overset{\|}{C}}-O-\overset{O}{\overset{\|}{C}}-CH_3$;

(31) 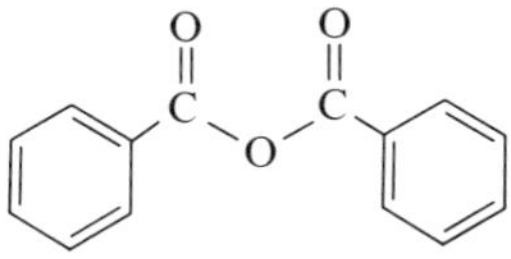 (32) 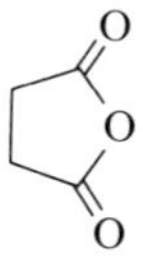 (33)

A239 (**B**). Compare with the CH_2 vibrations (**F42**).

F241 Would you expect acetic anhydride to absorb strongly below 1300 cm^{-1}?

A240 (**31**) 1815, 1750; (**32**) 1775, 1720; (**33**) 1865, 1780 cm^{-1}. Conjugation lowers frequency: 5-ring C=O absorbs at higher frequency than 6-ring/open chain.

Incidentally, another useful distinguishing feature comes from the relative intensities of the two bands. In the case of anhydrides with five-membered rings (**33**) the lower frequency component of the doublet is very much stronger than the higher frequency. With six-rings the difference is much less marked while for open chain anhydrides (**31** and **32**) the higher frequency component is slightly stronger than the lower.

F242 Unlike acids, esters, acid chlorides and acid anhydrides, amides absorb at lower frequency than analogous methyl ketones. The resonance effect involving the nitrogen lone-pair electrons outweighs the negative inductive effect partly because the electronegativity difference between nitrogen and carbon is less than that of oxygen or chlorine and carbon.

Without consulting tables, place the following compounds in order of increasing carbonyl frequency.

CH_3COCH_3 CH_3COOCH_3 CH_3CHO CH_3CONH_2 CH_3COCl

A241 Yes. The C—O stretching vibration would absorb strongly.

F243 Hydrogen atoms attached to nitrogen can be hydrogen-bonded to oxygen and other nitrogen atoms. Accordingly the carbonyl stretching frequency of an amide will be *lower*/*higher* in the solid phase or concentrated solution than in dilute solution.

A242 CH_3CONH_2, CH_3COCH_3, CH_3CHO, CH_3COOCH_3, CH_3COCl.

F244 Which of the following amides will have carbonyl stretching frequencies which, when measured in solution, are dependent upon concentration?

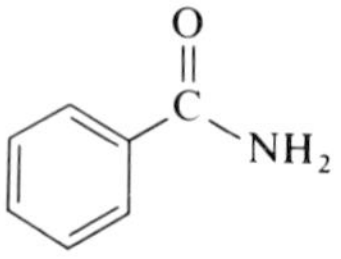

(34)

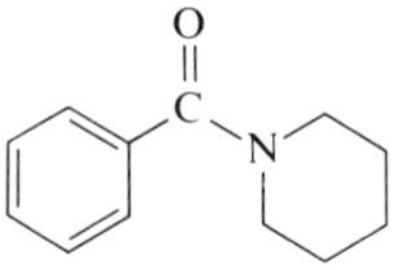

(35)

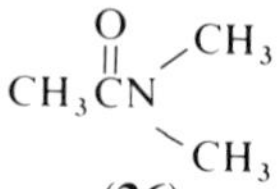

$CH_3CON(CH_3)_2$

(36)

$CH_3-C(=O)-NH-CH_2CH_3$

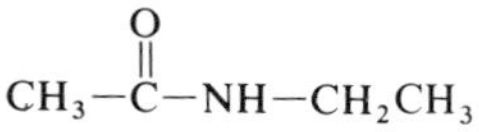

(37)

A243 lower. Note the analogy with carboxylic acids

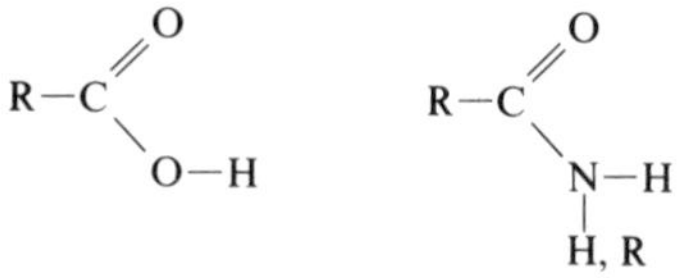

$R-C(=O)-O-H$ $R-C(=O)-N(H)-H, R$

F245 Propionamide, $CH_3CH_2CONH_2$, **in dilute solution** absorbs near 1690 cm^{-1}. Knowing that alkyl groups are electron donating relative to hydrogen atoms, would you expect a dilute solution of *N*-dimethylpropionamide to absorb at higher or lower frequency?

A244 (**34**) and (**37**). These are the only compounds with hydrogen attached to the nitrogen atom and hence likely to participate in hydrogen-bonding.

F246 When an alkenic or aryl group is attached directly to the nitrogen atom of an amide, the nitrogen lone-pair electrons can be delocalized away from the carbonyl group as well as towards it, i.e. the carbonyl group has some competition for the lone pair.

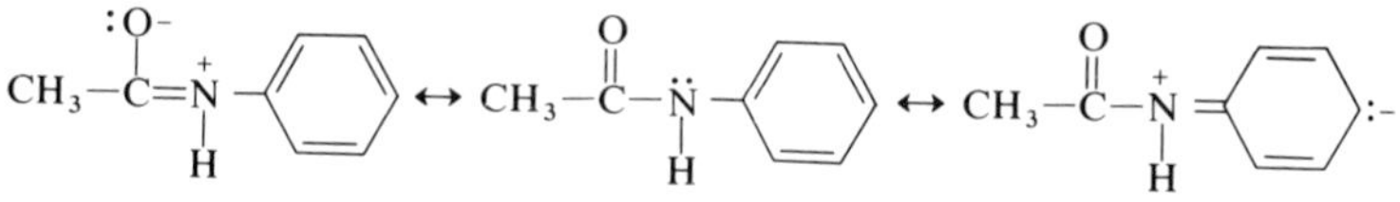

$CH_3-C(-\ddot{O}:^-)=\overset{+}{N}(H)-C_6H_5 \leftrightarrow CH_3-C(=O)-\ddot{N}(H)-C_6H_5 \leftrightarrow CH_3-C(=O)-\overset{+}{N}(H)=C_6H_5:^-$

Assign each of the following frequencies to one of the compounds given below, assuming all were measured under comparable conditions. 1650, 1660, 1685 cm^{-1}.

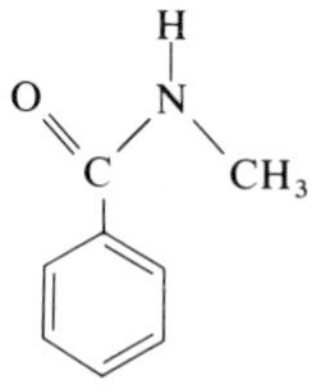

(38) $C_6H_5-CO-NH-CH_3$

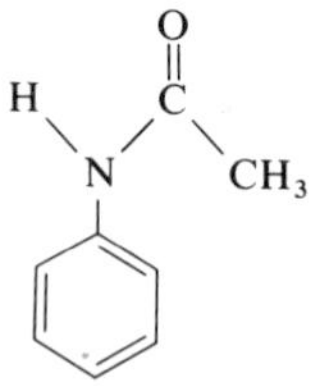

(39) $C_6H_5-NH-CO-CH_3$

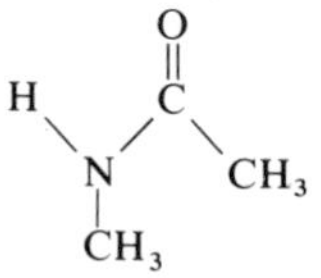

(40) $CH_3-NH-CO-CH_3$

A245 lower. As both compounds were measured in dilute solution we can ignore any hydrogen-bonding effect in propionamide. Donation of electrons to the nitrogen atom facilitates release of a lone-pair for resonance, so the contribution from the canonical form with C—O character increases.

$$(H_3C)_2N{-}C({=}O){-}CH_2{-}CH_3 \leftrightarrow (H_3C)_2\overset{+}{N}{=}C(\overset{-}{O}){-}CH_2{-}CH_3$$

F247 Using dilute solutions in non-polar solvents, the distinction between primary, secondary and tertiary amides can be made by examination of the high frequency region of the spectrum. Assign each of the partial spectra given below (on a larger wavenumber scale than usual) to one of the following compounds

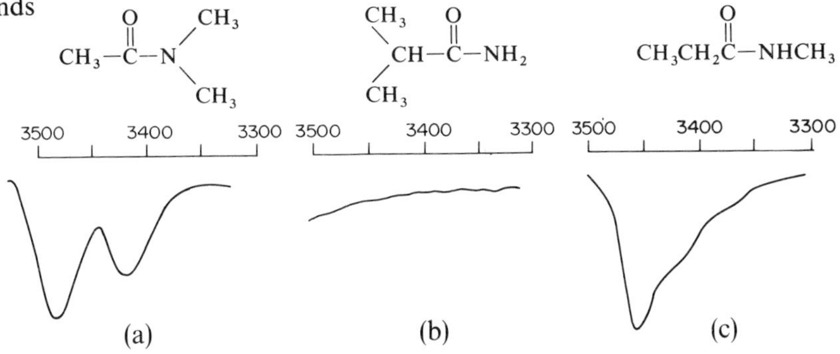

A246 **(38)** 1650; **(39)** 1685; **(40)** 1660 cm^{-1}.
All three are secondary amides and as they are measured under comparable conditions we can tentatively ignore hydrogen bonding. **(40)** may be regarded as the 'standard'. The conjugation with the ring in **(38)** lowers the carbonyl frequency. The carbonyl group in **(39)** is not conjugated with the aromatic ring and conjugation with the nitrogen lone-pair is less effective than in **(40)** because of sharing with the aromatic system. Thus **(39)** has the highest carbonyl frequency.

F248 Hydrogen-bonding, which occurs in the solid state and concentrated solutions often leads to multiple NH absorptions at relatively *low/high* frequency.

A247

(a) $(CH_3)_2CH{-}C({=}O){-}NH_2$

(b) $CH_3{-}C({=}O){-}N(CH_3)_2$

(c) $CH_3CH_2C({=}O){-}NHCH_3$

We are, of course, looking at NH stretching absorptions. The tertiary amide *N*-dimethylacetamide has no NH bond so CH absorption is the only high frequency feature in the spectrum. The primary amide α-methylpropionamide has symmetric and antisymmetric NH stretching vibrations which gives rise to a doublet above 3200 cm^{-1}.

F249 If you were given the spectrum of a sample of an amide made up as a Nujol mull which of the following would you be able to identify with certainty?

(a) primary amide
(b) secondary amide
(c) tertiary amide

A248 low. Hydrogen bonding increases the length of the N—H bonds involved.

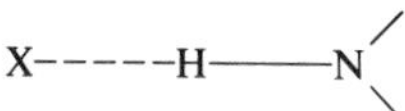

F250 Primary and secondary amides also show characteristic absorption (of moderate to weak intensity) between 1510 and 1650 cm^{-1}. This is due to NH in-plane bending vibrations. Assign each of the lettered absorption bands in the spectrum of propionamide $CH_3CH_2CONH_2$, (Fig. 4.8), to one of the vibrations listed below.

(a) NH in-plane bending
(b) NH out-of-plane bending
(c) CH stretching vibration
(d) C=O stretching vibration
(e) NH stretching vibration
(f) skeletal vibration

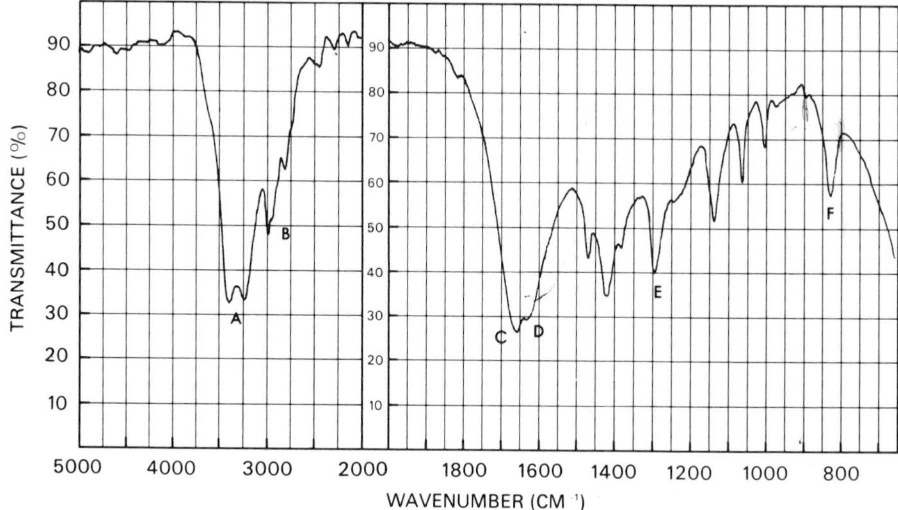

Fig. 4.8. Spectrum of propionamide (solid in KBr).

A249 tertiary. The absence of any absorption above 3100 is unambiguous. (If the compound is not already known to be an amide, identification of tertiary amides is much less certain since the only evidence available is the low carbonyl frequency.)

F251 Assign each of the following compounds to one of Figs. 4.9—4.14. The sampling technique and the phase of the sample is stated with each spectrum.

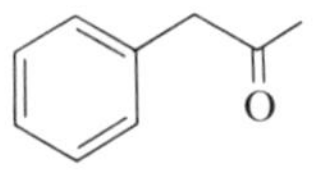

O (**41**)

CHO CH_3 (**42**)

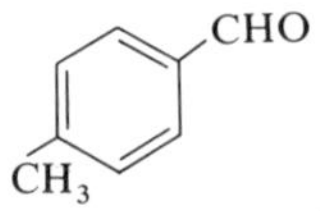

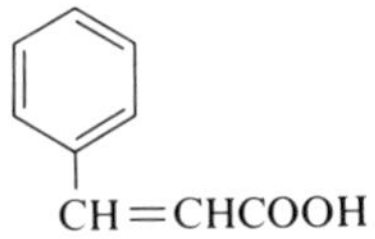

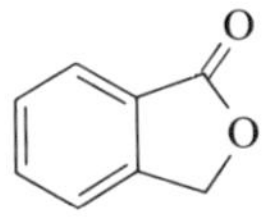

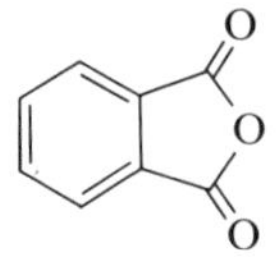

CH=CHCOOH (**43**)

O O (**44**)

O O O (**45**)

NH_2 O (**46**)

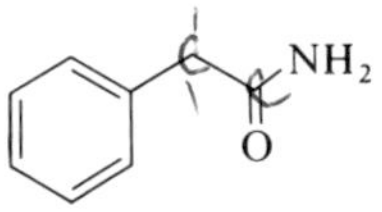

A250 (a) D; (b) F; (c) B; (d) C; (e) A; (f) E.

A251 (**41**) 4.10; (**42**) 4.14; (**43**) 4.13; (**44**) 4.11; (**45**) 4.12; (**46**) 4.9.

There are several ways of tackling a problem like this. One is to look first for some distinguishing feature in one of the structures and then to scan through the spectra until one is found containing a band characteristic of this feature. For example, the primary amide (**46**) should show a doublet near 3300 cm^{-1} whilst the anhydride (**45**) should show a doublet at the high frequency end of the carbonyl region. The acid (**43**) should show a broad OH band centred at about 3000 cm^{-1} and the aldehyde (**42**) a doublet near 2700 cm^{-1}. Hence these four compounds correspond to Figs. 4.9, 4.12, 4.13 and 4.14 respectively. There are other features such as the precise frequency of the C=O bond which would confirm this assignment. We can also expect to distinguish (**41**) and (**44**) by the C=O frequency which should be about 1735 for (**44**) where the decrease due to conjugation is balanced by the increase due to the five-membered ring. Hence the assignment of 4.11 to (**44**) and 4.10 to (**41**). Note also the strong absorptions in the 1000—1100 cm^{-1} region in 4.11 associated with the C—O bonds in (**44**) (and absent in (**41**)).

Other Nitrogen-containing Groups

F252 Cyanides and isocyanides both contain formal triple bonds.

$$R{-}C{\equiv}N \qquad R{-}\overset{+}{N}{\equiv}\overset{-}{C}$$

As such, they should absorb in a unique region of the spectrum, making recognition of the group by infrared spectroscopy a straightforward matter. Figure 4.15 is the spectrum of ethyl cyanoacetate. Write down the frequency of the 'unique' band and the vibration responsible.

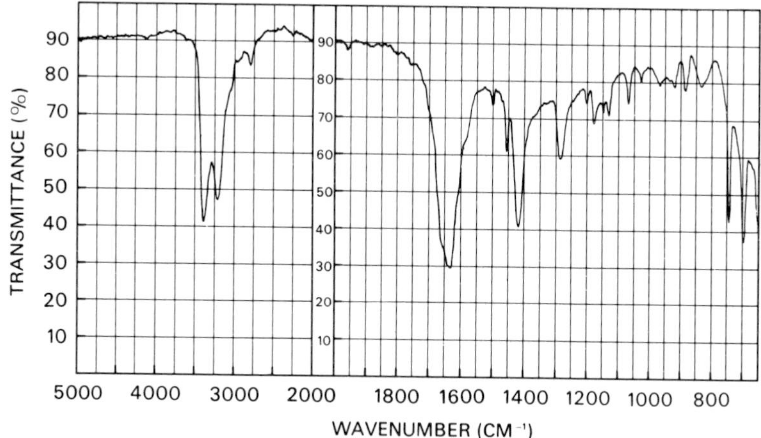

Fig. 4.9. Solid, split mull (**F251**).*

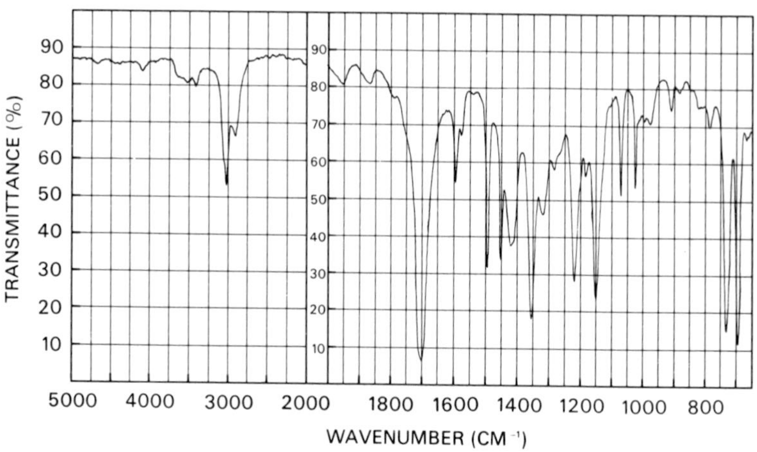

Fig. 4.10. Liquid, capillary film (**F251**).

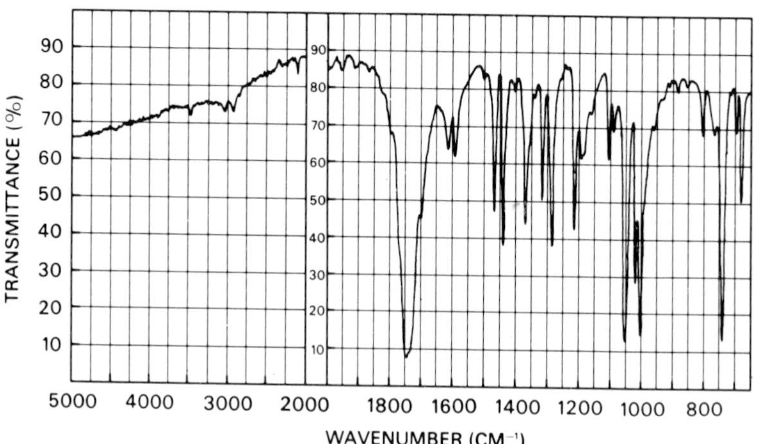

Fig. 4.11. Solid, split mull (**F251**).*

* Hexachlorobutadiene mull used for the 1300—1600 and 2000—4000 cm^{-1} régions. Nujol mull used for remainder of the spectrum.

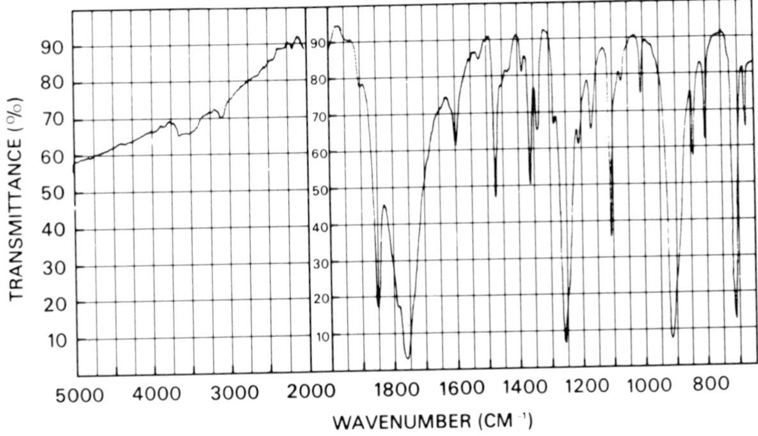

Fig. 4.12. Solid, 1% disc in KBr (**F251**).

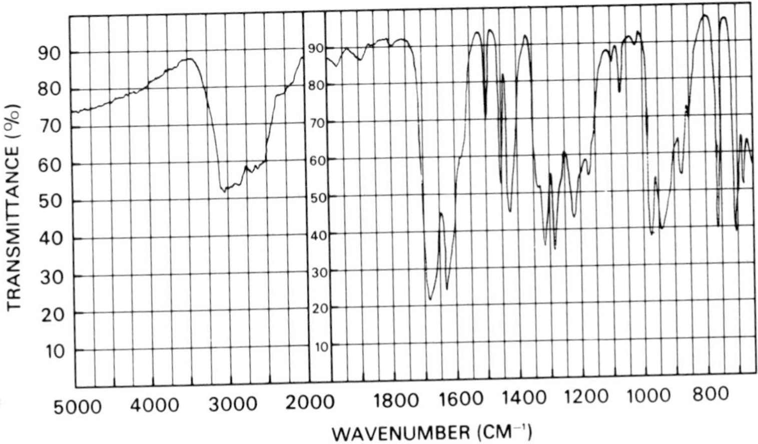

Fig. 4.13. Solid, split mull (**F251**).

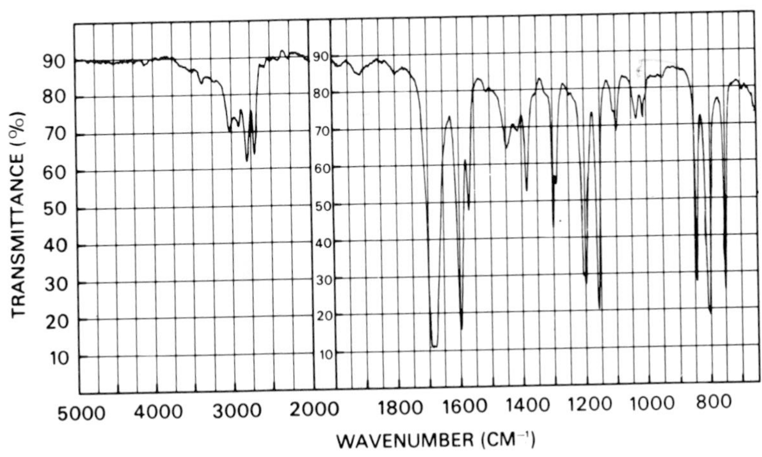

Fig. 4.14. Liquid, capillary film (**F251**).

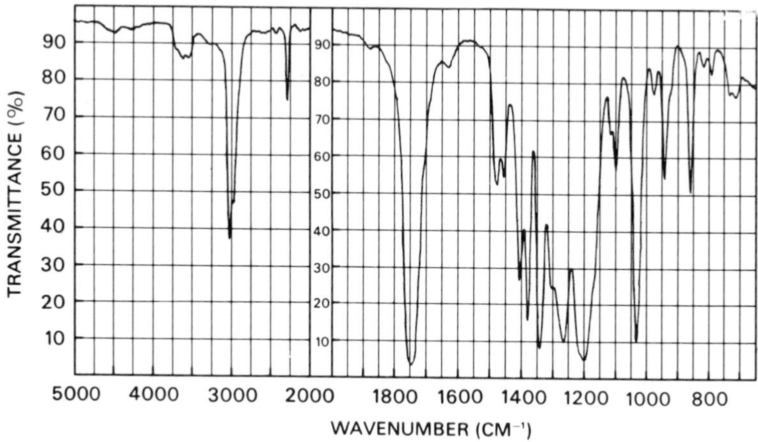

Fig. 4.15. Spectrum of ethyl cyanoacetate (**F252**).

F253 Cyanides and isocyanides can be distinguished by the frequency of the C≡N stretching vibration.

alkyl cyanides R—C≡N	alkyl isocyanides $R—\overset{+}{N}≡\overset{-}{C}$
2240—2270 cm^{-1}	2100—2190 cm^{-1}

However, isocyanates, R—N=C=O absorb in the same region as cyanides (compare the high frequency observed for the C=C=C group in allenes which absorbs in the same region as acetylenes—see **A187**). They can be distinguished, however, by the fact that isocyanates have much stronger absorption bands than cyanides. Assign each of the following compounds to one of the partial spectra given below.

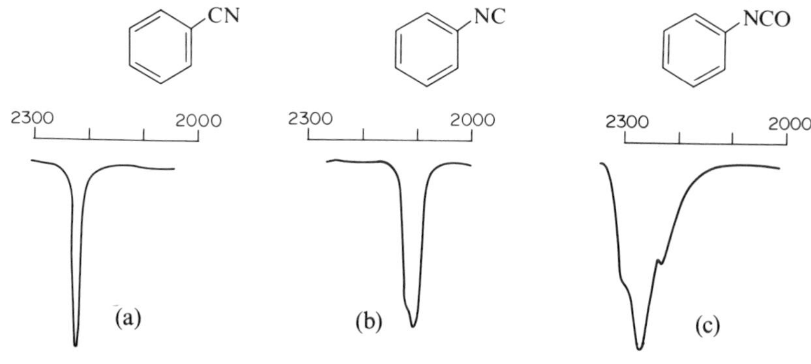

A252 2270 cm^{-1}; C≡N stretch. It is special because very little else absorbs in the 2000—2500 cm^{-1} region.

F254 C=N and N=N are two other common multiple bonds containing nitrogen. By now you should be able to speculate on the region where absorption due to their stretching vibrations may be found.

Select one from the following alternatives.

(a) 2600—3400; (b) 2000—2600; (c) 1800—2000;
(d) 1400—1700; (e) 800—1400 cm^{-1}.

A253 (a) 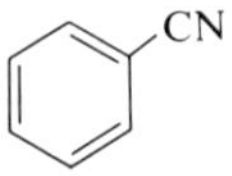 [benzene ring with CN] (b) 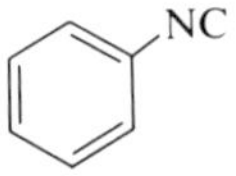 [benzene ring with NC] (c) 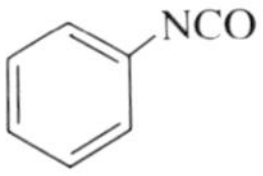 [benzene ring with NCO]

F255 Which of the following would you expect to be true?

(a) Both C=N and N=N bands are very weak;
(b) C=N is much stronger than N=N;
(c) N=N is much stronger than C=N;
(d) both are very strong absorptions?

A254 (d). C=N frequencies, like C=O are strongly influenced by neighbouring groups and may occur anywhere with the range stated. N=N usually occurs near 1570 cm^{-1}.

F256 The nitro group $-NO_2$ is best regarded as a resonance hybrid between two equivalent forms.

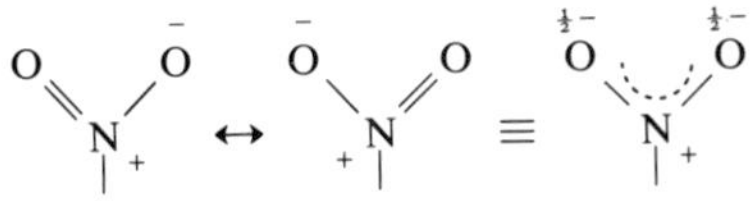

$$O{=}\overset{+}{N}{-}O^{-} \leftrightarrow {}^{-}O{-}\overset{+}{N}{=}O \equiv O^{\frac{1}{2}-}\cdots\overset{+}{N}\cdots O^{\frac{1}{2}-}$$

Would you expect the N—O stretching absorption to appear as

(a) one very intense band
or (b) two very weak bands
or (c) two very strong bands
or (d) one very weak band?

A255 (b). Recall that the strength of an infrared absorption band is related to the change in dipole moment during the vibration. The infrared spectrum of *trans*-azobenzene

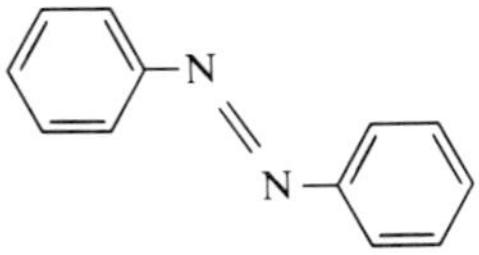

[structure: Ph–N=N–Ph]

shows no absorption due to the N=N stretch.
(Indeed the N=N stretching band is so weak that it is seldom observed in infrared spectra.)

F257 Saturated aliphatic nitro-compounds absorb strongly near 1550 cm^{-1} and 1370 cm^{-1}. Conjugation lowers the frequency to an extent determined by the electron donating power of the unsaturated group. Assign each of the following pairs of frequencies to one of the compounds given below.

(a) 1480 and 1319; (b) 1560 and 1360; (c) 1530 and 1355 cm^{-1}.

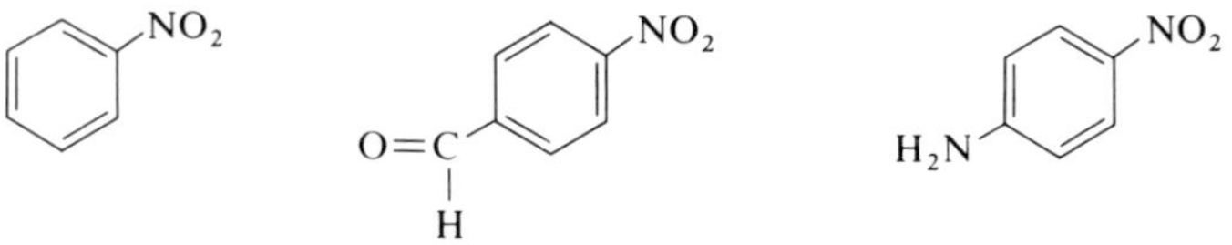

A256 (c). The situation is entirely analogous to the carboxylate anion. Two equivalent bonds to the same atom means that we can have asymmetric and symmetric stretching modes, both of which will be very strong because of the polar nature of the group.

A257

(a) 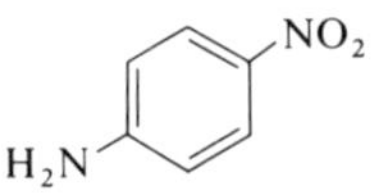 (b) 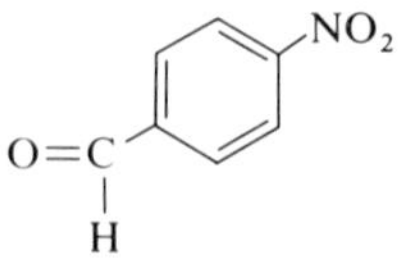 (c)

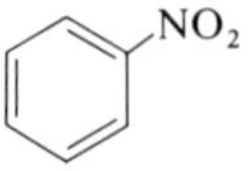

An amino group ($-NH_2$) is electron donating; a carbonyl group ($>\overset{\delta+}{C}=\overset{\delta-}{O}$) is electron withdrawing.

Sulphur- and Halogen-containing Compounds

F258 The S—H bond has a much lower stretching frequency (2520—2600 cm^{-1}) than C—H, O—H or N—H. The band is intrinsically very weak and absence of absorption in this region does not necessarily indicate the absence of S—H in the molecule. The S=O stretching frequency of sulphoxides R—SO—R, occurs near 1090 cm^{-1} and the vibration absorbs infrared strongly. Assign each of the spectra in Fig. 4.16 to one of the following compounds:

C_2H_5SH; $C_2H_5SC_2H_5$; $C_2H_5SOC_2H_5$

F259 Like sulphoxides, sulphones ($R-SO_2-R$) absorb strongly below 1400 cm^{-1}. In addition to difference in frequency, there is an easy way to distinguish absorption due to the sulphone group from that due to the sulphoxide group. What is it?

A258 A, C_2H_5SH; B, $C_2H_5SC_2H_5$; C, $C_2H_5SOC_2H_5$.

F260 The S=O stretching frequencies of sulphonamides ($R-SO_2-NH_2$) have approximately the same values as those of sulphones. It seems, therefore, that

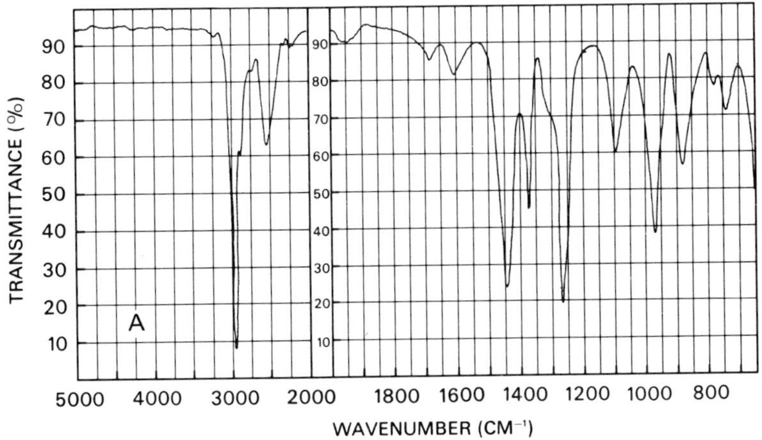

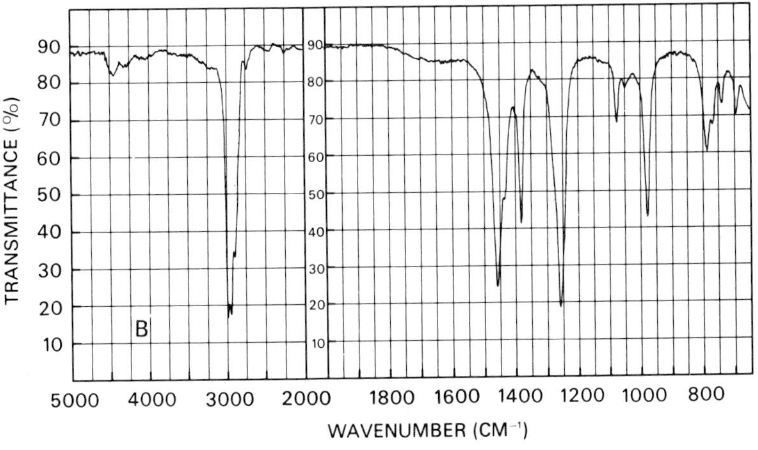

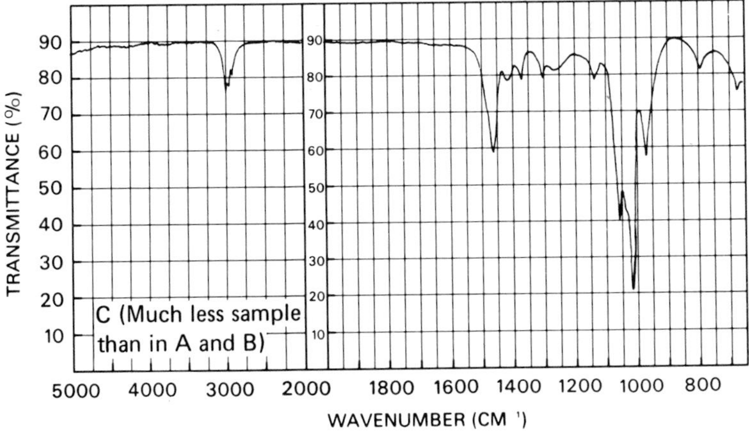

Fig. 4.16. Spectra of sulphur compounds (**F258**).

the negative inductive pull of the nitrogen atom is counterbalanced by the resonance contribution (cf. ketones, RCOC<, and amides, $RCONH_2$).

$$R{-}S(=O)_2{\rightarrow}NH_2 \qquad R{-}S(=O)(-O^-){=}N^+H_2 \leftrightarrow R{-}S(=O)(-O^-){=}N^+H_2$$

In view of this, which of the following pairs of ranges would you predict to be the most likely for sulphonyl chlorides, RSO_2Cl?

(a) 1050—1100 and 1250—1300 cm^{-1}; (b) 1180—1210 and 1380—1410 cm^{-1}.

A259 The sulphone group would have two S=O stretching frequencies corresponding to asymmetric and symmetric modes.

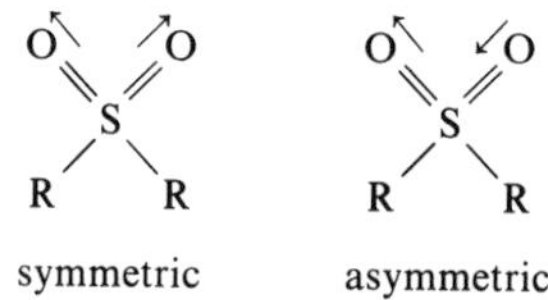

symmetric asymmetric

Actual frequencies are: 1120—1160 cm^{-1} and 1300—1350 cm^{-1}. Both are strongly absorbing.

F261 Halogen containing compounds usually absorb strongly at the frequency of the carbon–halogen stretching vibration. This covers the range 600—850 cm^{-1} in the case of organochlorine compounds.

Which of the following statements is/are true?

(a) C—Br stretches occur between 1000 and 1100 cm^{-1}.
(b) C—I stretches occur below the frequency range of many routine instruments.
(c) C—F stretches occur in the range 1000—1400 cm^{-1}.
(d) C—F stretches occur below the frequency range of many routine instruments.

A260 (b) 1180—1210 and 1380—1410 cm^{-1}. The situation is entirely analogous to the comparison between carboxylic amides and acid chlorides where the effect of the chlorine atom is to increase the frequency of the neighbouring carbonyl group due to the electron withdrawal.

F262 Most of the absorption characteristics of heteroaromatic compounds such as pyrrole, furan and quinoline can be recognized in terms of being similar to their hydrocarbon analogues on the one hand and related to group frequencies involving heteroatoms on the other.

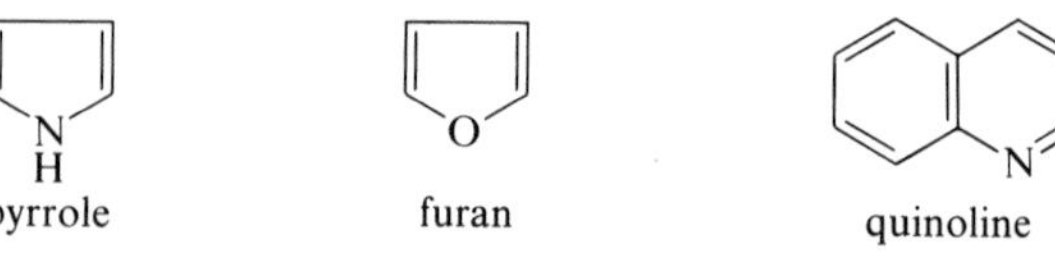

pyrrole furan quinoline

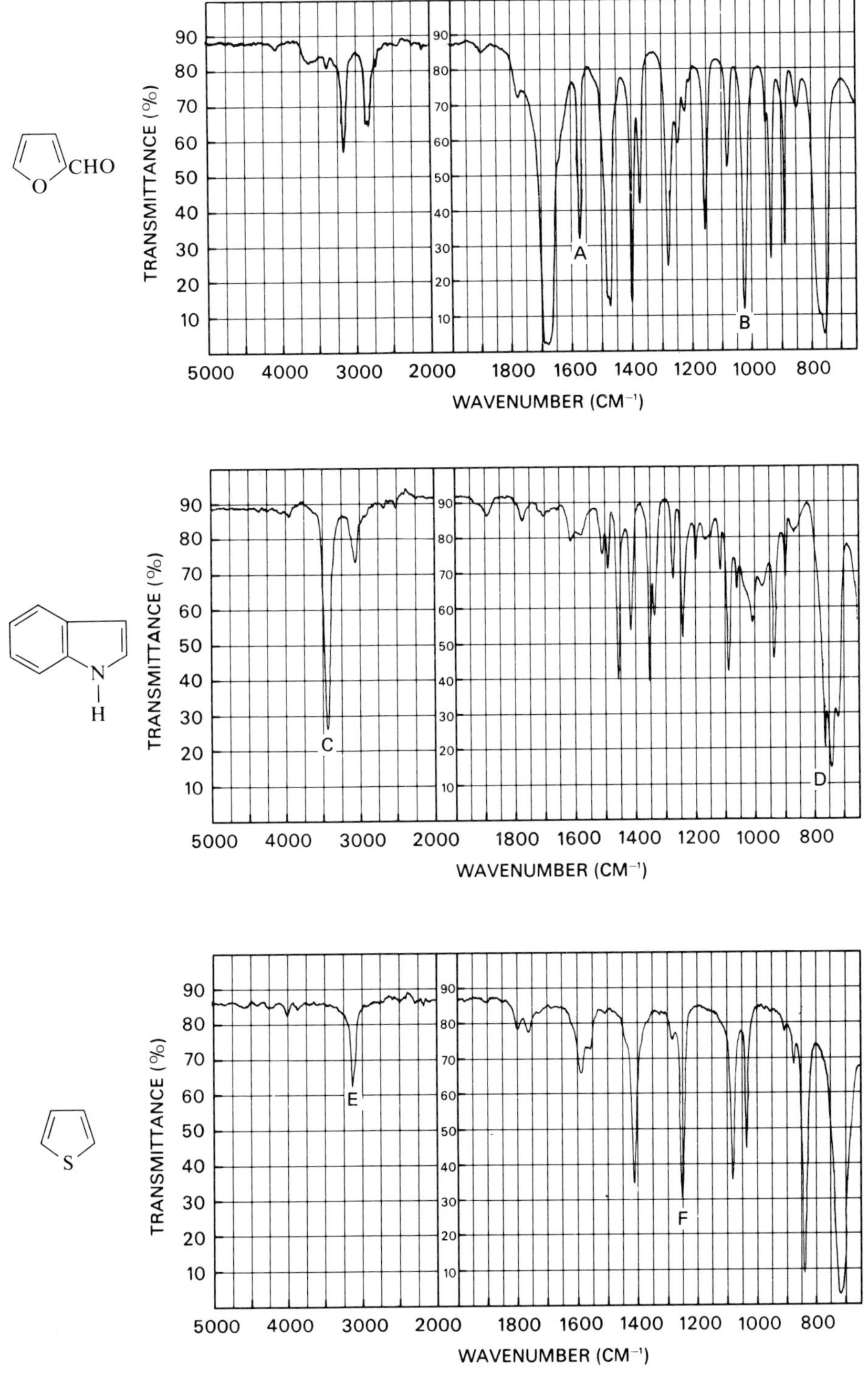

CHO
O
TRANSMITTANCE (%)
90
80
70
60
50
40
30
20
10
A
B
5000
4000
3000
2000
1800
1600
1400
1200
1000
800
WAVENUMBER (CM⁻¹)
N
H
C
D
E
F
S

Fig. 4.17

Assign each of the labelled bands in the spectra of Fig. 4.17 to one of the following vibrations:

NH stretch;
a ring stretching vibration involving multiple bonds;
CH stretch;
CH out-of-plane bend;
skeletal vibrations involving C—O;
skeletal vibrations involving in-plane CH bend.

A261 (b) and (c) are true. C—Br stretches also occur below the frequency range of many routine instruments.

A262 A, ring stretching vibration involving multiple bonds; B, skeletal vibrations involving C—O; C, NH stretch; D, CH out-of-plane bend; E, CH stretch; F, skeletal vibrations involving in-plane CH bend.

Revision Summary

Saturated hydrocarbons

Useful absorption frequencies are restricted to C—H bending and stretching. Methyl and methylene vibrations can only be distinguished under high resolution. Compounds containing a *gem*-dimethyl group often show a characteristic doublet near 1375 cm^{-1}.

Unsaturated hydrocarbons

Vibrations which give rise to absorptions are:

(i) out-of-plane CH bend;
(ii) in-plane CH bend, (not generally useful);
(iii) >C=C< stretch;
(iv) =C—H stretch.

These are observed for both alkenes and benzenoid hydrocarbons and in both cases the precise frequencies of (i) and (iii) can be structurally revealing. The two classes of compound can frequently be distinguished by the fact that benzenoid ring C=C stretching often results in moderate absorption near 1500 cm^{-1}. This is never observed with alkenes.

Alcohols and phenols

These are characterized by strong absorptions above 3000 cm^{-1} and in the fingerprint region corresponding to O—H and C—O stretching, respectively. The frequency of the former is dependent on the extent of hydrogen bonding under the conditions of measurement and can give rise to very broad bands. Hydrogen bonding is minimized by using very dilute solutions in non-polar solvents.

Amines

N—H stretches of primary and secondary amines occur within the broad O—H region and are similarly affected by hydrogen bonding. O—H and N—H stretching absorptions are often distinguishable by the sharpness of the latter and, in the case of a primary amine, the appearance of a doublet. Another characteristic absorption of primary amines is the N—H bend, usually observed in the 1580—1650 cm^{-1} region.

Ethers

Once it is known that a compound contains oxygen, establishing whether or not it is an ether is best done by elimination of C=O and O—H. The only positive evidence in an infrared spectrum would be the presence of strong absorptions in the 1000—1250 cm^{-1} region associated with the C—O bond.

Ketones and aldehydes

These can be distinguished by the unique C—H stretching frequency of the CHO group. The most diagnostic and informative feature of both groups is the frequency of the C=O stretching vibration. Hydrogen bonding, conjugation and (in ketones) C—CO—C bond angle are critical features here.

Carboxylic acids, esters and lactones

The carbonyl stretching frequencies of carboxylic acids are usually observed at a relatively low frequency because, under most conditions of measurement, they exist as hydrogen-bonded dimers. This feature also results in a characteristic very broad O—H stretching band which 'straddles' the CH region. Carboxylate anions show two strong carbonyl absorptions corresponding to symmetric and antisymmetric modes. The C=O stretching frequency of esters is higher than analogous ketones but the regions for each group do overlap and in cases of doubt an ester is confirmed by the presence of a strong C—O band in the fingerprint region. Cyclic esters (lactones) vary in frequency in a precisely analogous manner to cyclic ketones. Unsaturation can raise or lower ester carbonyl frequencies depending on the atom to which it is attached.

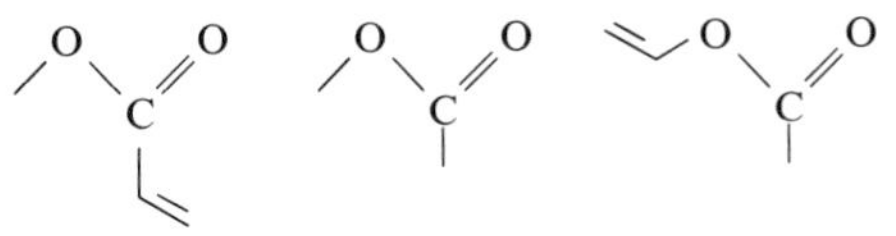

lower normal higher

Carboxylic acid halides, anhydrides and amides

The most characteristic feature in the infrared spectra of acid halides is the very high C=O stretching frequency that is always present. Anhydrides also absorb at the high end of the carbonyl range. They are easily recognized by the fact that the absorption usually appears as a doublet. Amides, on the other hand, show some of the lowest carbonyl stretching frequencies normally encountered. The NH and NH_2 groups of secondary and primary amides give rise to very similar absorption features to corresponding amines. Both NH and C=O stretching absorptions are dependent on the extent of hydrogen bonding. The effect on the C=O frequency of unsaturation is similar to that for esters.

Cyanides, isocyanides and isocyanates

The C≡N stretch near 2250 cm^{-1} is the characteristic feature here. Cyanides can be distinguished from isocyanides by frequency and from isocyanates by strength of absorption.

Other groups

The nitro group is recognizable in the infrared by two strong absorptions near 1550 and 1370 cm^{-1}. N=N and C=N generally give rise to weak or moderate strength bands near 1600 cm^{-1} corresponding to stretching of the bond. Of the commonly occurring groups containing sulphur, RS—H is recognizable by the weak SH stretching absorption near 2500 cm^{-1}, whilst strong absorptions at characteristic frequencies are diagnostic of groups of the type R—SO—R and R—SO_2—R. The C—Hal bond of halogen containing compounds produces very strong absorptions but except for fluorine, they are usually observed near the lower frequency limit of routine spectrophotometers.

Return now to the multiple-choice test at the beginning of Part 4 and mark it as before. Then consult the remarks below.

Less than 6 Consult your instructor as to whether you should go on or seek extra tuition. If you are using the programme for private study, re-read the frames pertinent to the questions you answered incorrectly and, if possible, consult the references for further reading. When you are satisfied you fully understand the material, read the paragraph at the end of this Part and start the next Part.

6 or more Run over the sections of this part covering the material concerned with the questions you answered incorrectly. Consult your teacher as to whether you should spend some time on the exercises and bibliography which follows.

Questions

1. Write an account of the influence of hydrogen-bonding on the infrared spectra of organic compounds.

2. How would you distinguish between the following pairs of compounds by infrared spectroscopy?

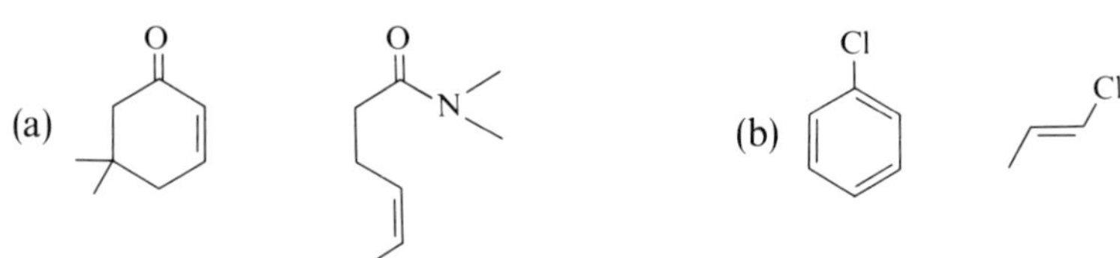

(c) 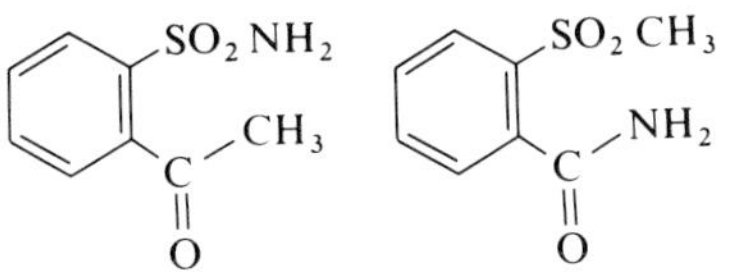 (d)

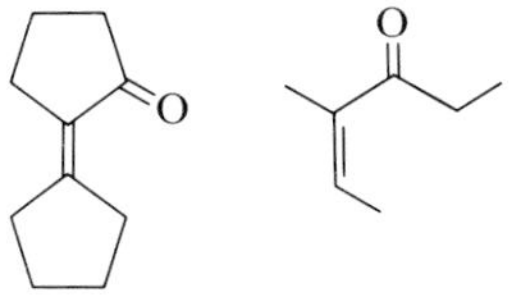

(e) 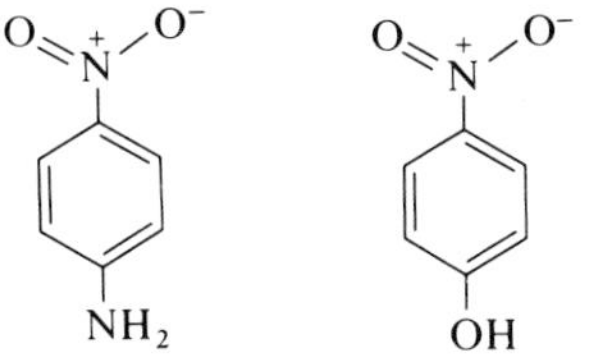 (f)

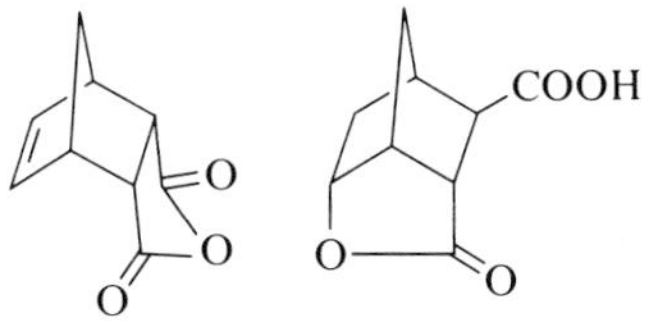

3. Predict the characteristic features of the infrared spectra of the following compounds.

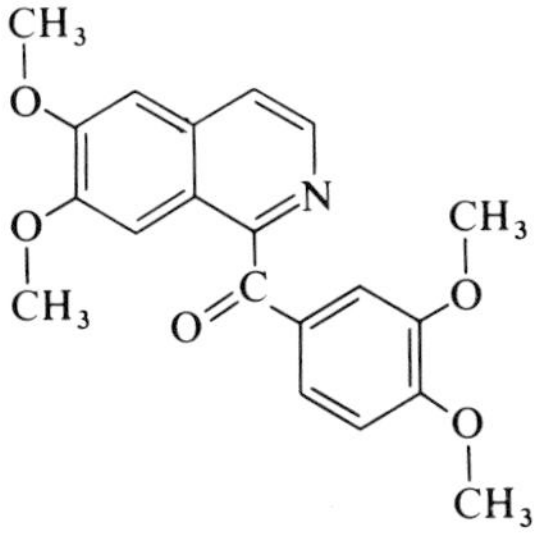

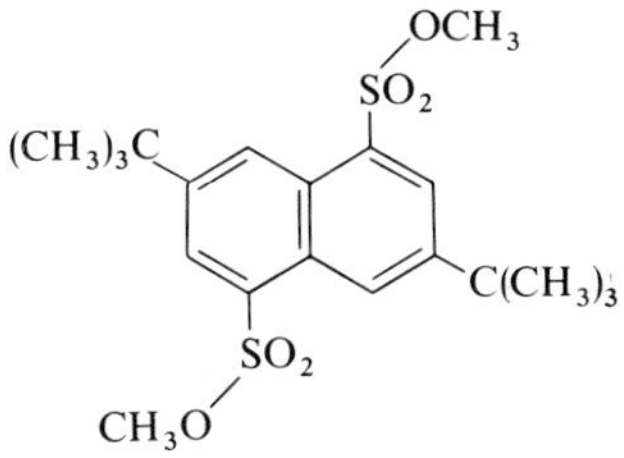

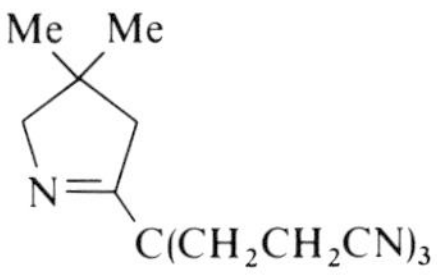

$CH_3CH(OH)CH(NH_2)CO_2H$

Further Reading

J. R. Dyer, *Applications of Absorption Spectroscopy of Organic Compounds*, Prentice Hall, Englewood Cliffs, N.J., 1965, Chapter 3.
R. M. Silverstein and G. C. Bassler, *Spectrometric Identification of Organic Compounds*, John Wiley & Sons, New York, 1967, Chapter 3.
D. H. Williams and I. Fleming, *Spectroscopic Methods in Organic Chemistry*, McGraw-Hill, New York, 1966, Chapter 3.
A. D. Cross and R. A. Jones, *An Introduction to Practical Infrared Spectroscopy*, Butterworths, London, 1969.
Y. Yukawa, *Handbook of Organic Structural Analysis*, Benjamin, New York, 1965.
A. Weissberger, *Techniques of Organic Chemistry*, Volume II, Part 1, John Wiley and Sons, New York, 1963.
D. J. Pasto and C. R. Johnson, *Organic Structure Determination*, Prentice Hall, Englewood Cliffs, N.J., 1969, Chapters 4 and 12.
L. J. Bellamy, *Infrared Spectra of Complex Molecules*, 2nd Edn., Methuen, London, 1958.
L. J. Bellamy, *Advances in Infrared Group Frequencies*, Methuen, London, 1968.
M. Avram and Gh. D. Mateescu, *Infrared Spectroscopy, Applications in Organic Chemistry*, Wiley-Interscience, New York, 1972.

We have now covered most of the common features you are likely to meet in the infrared spectra of organic compounds. Indeed, in some cases you have met them more than once but have considered them from different points of view. This is perhaps the most appropriate point to mention briefly the infrared absorption characteristics of compounds which are entirely inorganic. Knowledge of these characteristics is useful to the organic chemist particularly in syntheses where an unknown 'product' could well be derived from an inorganic reagent. Fortunately this situation is easily recognised in most cases since the physical properties of inorganic compounds are quite different from those of organic compounds. An inorganic material is likely to be a solid of high melting point which is insoluble in organic solvents though it may dissolve in water or aqueous solutions of acids or bases. The compounds most likely to be encountered in synthetic organic work are:

(a) oxides and hydroxides of metals (including silicon);
(b) metallic salts of oxy acids (and as phosphates, sulphates etc.);
(c) metallic halides;
(d) ammonium salts.

The oxygen containing compounds generally show a broad region of intense and relatively featureless absorption in the range 800—1500 cm^{-1} due to vibrations involving metal–oxygen bonds of oxyanions. These compounds may also contain water of crystallization or adsorbed water which gives rise to a strong broad band at 3300 cm^{-1} **together with** a weaker band at 1650 cm^{-1} and 'end absorption' starting at about 700 cm^{-1}.

Metallic halides do not absorb at all in the infrared region above 700 cm^{-1} unless they contain water of crystallization (many do).

Ammonium salts are characterised by a sharp band due to the NH_4^+ ion at 1400 cm^{-1} together with a band near the water band at 3100—3200 cm^{-1}.

Purely inorganic materials of course do not give rise to the bands most characteristic of all organic compounds, C—H stretching and bending unless they contain traces of organic impurities.

You should now have the basic tools to enable you to use the full value of infrared spectroscopy in structure elucidation of organic compounds. All you need now is that vital ingredient—experience. We hope that by using the programme method you have already acquired some! The final part of the programme, however, is concerned solely with this aspect.

PART 5
PROBLEMS

Aim

By the time the student has reached this part of the programme he should be able to obtain structural information from infrared spectra empirically using correlation charts and tables. He should also be aware of the factors which influence the precise position of characteristic frequencies within their quoted range and, as a consequence, be able to obtain more information about the structure of an unknown than would be possible by a mere identification of functional groups. He will also have developed some technique in the interpretation of infrared spectra and it is the principal aim of Part 5 to present an opportunity whereby this may be improved.

All the prior knowledge required for Part 5 should have been acquired in the previous parts and no new terms or concepts will be introduced. The opportunity will be taken however to revise and test material presented earlier and also to introduce new information relating to special correlations as they arise.

Objective for Part 5

When you have completed Part 5 you should be able to interpret the infrared spectrum of a known substance, or use the infrared spectrum of an unknown substance to formulate plausible alternatives for its structure. You should be able to achieve this more rapidly and effectively than before.

PART 5: PROBLEMS

Introduction

In this final part of the programme you will attempt to identify a series of compounds from their spectra making use of the principles which should now be familiar from your reading of the earlier parts. Only on very rare occasions is an infrared spectrum sufficient by itself to enable a compound to be identified, and in the problems which follow supplementary information will be supplied to allow the identification to be completed. This may be a molecular formula, molecular weight or some other property. In the first problem you will be guided step by step in your interpretation of the spectrum. In later problems the extent of this guidance will be progressively reduced until finally you will undertake the interpretation entirely on your own.

You should make use of the correlation charts throughout (Charts 2—4) though you will probably find that you will need Chart 2 less and less as you become more familiar with the general position of key characteristic frequencies. All the charts are on the sheet inserted in the pocket at the back of the book. For those problems at the end of the programme where no guidance is given you may find it useful to refer back to the Revision Summary of Part 4 once you think you have established the class of compound in question. The answers to the questions in this part are frequently much longer and more detailed than in the earlier parts. They will also contain information relating to specific correlations which you are not expected to memorise. This information is however collected together for reference purposes in Chart 5. Unlike the previous parts of the programme there is no pre-test or post-test associated with Part 5 and for further problems you are referred to the many excellent books of spectroscopic problems given in the further reading section.

No hard and fast rules can be laid down for interpreting an infrared spectrum. There are certain guidelines however which have proved useful, especially when the molecular formula of the substance is known. Inspection of the spectrum and the molecular formula will reveal immediately the presence of some functional groups and the absence of others. Tentative assignments can be confirmed and new ones made by a more detailed examination of band frequencies and correlation charts. Subtraction of the elements already accounted for from the molecular formula will give a residual formula the nature of which might suggest other groupings which can be confirmed or eliminated by further examination of the spectrum. When this iterative process is complemented by information about the compound available from other sources the full structure or, at worst, a narrow range of possibilities can be deduced. The most important sources of additional structural information are other spectroscopic methods such as nuclear magnetic resonance and mass spectroscopy. In practice the origin of the material often suggests its identity.

But to begin with let us see how far we can get using the infrared spectrum alone.

Problem 1

Identification of a colourless liquid

F263 The compound has the spectrum shown in Fig. 5.1. First of all write down the frequency of the strongest band and the functional group responsible for it.

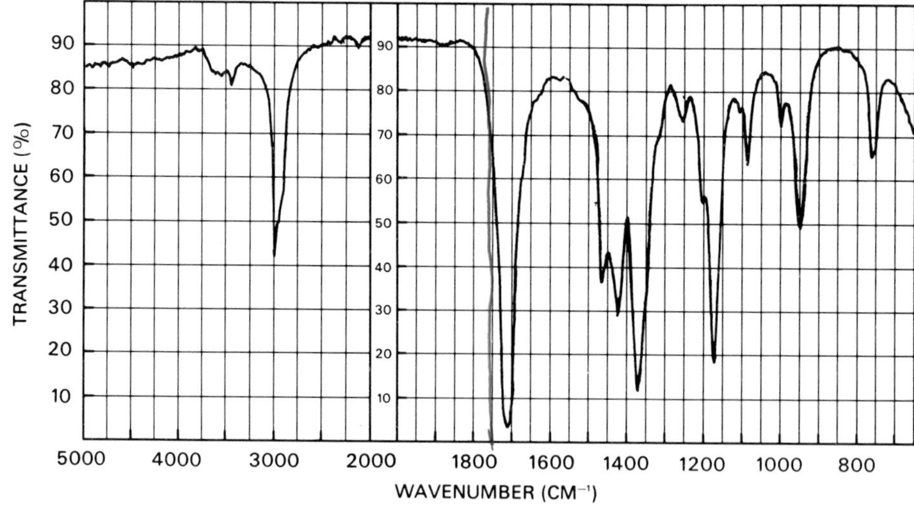

Fig. 5.1

F264 The presence or absence of a carbonyl group is perhaps the most certain piece of information to be obtained from an infrared spectrum. The carbonyl group occurs in a wide range of compounds and the next step is to determine what type of carbonyl compound we have.

Which of the following types are excluded by the fact that the absorption occurs above 1700 cm^{-1}?

aldehyde, open chain ketone, $\alpha\beta$-unsaturated ketone, ester, carboxylic acid, acid chloride, amide (see Chart 4).

- -

A263 1705 cm^{-1}; $>$C=O.

F265 Which of the following types are eliminated by the fact that the absorption occurs below 1750 cm^{-1}?

aldehyde, open chain ketone, carboxylic acid, acid chloride, anhydride, ester.

- -

A264 amide and $\alpha\beta$-unsaturated ketone are eliminated.

F266 We are thus left with the following possibilities:

aldehyde, open chain ketone, ester, carboxylic acid.

Although we could narrow down the possibilities further by making use of the precise frequency of the carbonyl band there is still some overlap in the ranges quoted for these four types. Besides this, before laying too much emphasis on the exact frequency of an absorption band, one should always take steps to check the calibration of the instrument on which the spectrum was run (see Part 2). Errors of up to 5 cm^{-1} are quite common with routine instruments in the 1700 cm^{-1} region. Finally, if the band is broad or unsymmetrical (as is often the case with C=O bands) it will be difficult to assess its centre more accurately than about $\pm 5\ cm^{-1}$. Fortunately, in the present case we do not need to rely on the precise frequency of the C=O band since three of the four remaining classes of compound will give rise to characteristic absorptions in other regions of the spectrum and these will enable us to decide between the possibilities.

Write down the type of vibration (apart from C=O stretching) which will give rise to a recognizable feature in the spectrum of compounds containing the groups

$$-CHO \qquad -CO_2H \qquad \text{and} \qquad -CO_2R.$$

A265 acid chloride and anhydride are eliminated.

F267 In what region of the spectrum shall we check whether or not we have a carboxylic acid?

Is the compound a carboxylic acid?

A266 CHO — C—H stretching
CO_2H — O—H stretching, and C—O stretching
CO_2R — C—O stretching

F268 In what region of the spectrum shall we check whether or not the compound is an ester?

Is the compound an ester?

A267 Above 2500 cm^{-1}. Because of hydrogen-bonding the OH stretching region of carboxylic acids gives rise to a very broad band extending from 2500 to 3500 cm^{-1}. There is no broad absorption of this type so the compound is not a carboxylic acid.

F269 In what region of the spectrum shall we check whether or not the compound is an aldehyde?

Is the compound an aldehyde?

A268 A **strong** band in the 1000—1300 cm^{-1} region would confirm the ester. In fact, there is a prominent band at 1170 cm^{-1} but this is not strong enough to be due to C—O. Even in quite large molecules the C—O band is stronger than the bands near 1400 cm^{-1} due to C—H bending. In fact its intensity is normally comparable with that of the C=O band.

F270 Thus it appears that the compound is a simple ketone. The carbonyl frequency is too high for an $\alpha\beta$-unsaturated ketone but this does not exclude the possibility of C=C links more remote from the carbonyl group. Look at the C—H stretching region. Which of the following interpretations of this region is correct?

(a) The compound contains saturated C—H bonds only;
(b) The compounds contains alkene C—H bonds only;
(c) The compound contains both aliphatic and alkene C—H bonds.

- - - - - - - - - - - - - - - - - - - -

A269 2650—2820 cm^{-1}. The CH absorption of the aldehyde group commonly appears as a doublet because of Fermi resonance (**F77**).

The compound is not an aldehyde. There is no absorption band between 2500 and 2850 cm^{-1}.

F271 All compounds containing aliphatic C—H groups show a doublet in the 2900—3000 cm^{-1} region of the spectrum though in the present case the separate components are less well-resolved than usual.

Which of the following statements correctly accounts for the appearance of just two bands in this region?

(a) One band is due to CH_3, the other to CH_2.
(b) All saturated CH groups give rise to absorption at the same frequency but this is split into a doublet by Fermi resonance with the overtone of a CH bending vibration near 1400 cm^{-1}.
(c) The doublet arises from the symmetrical and antisymmetrical vibrations of CH_2 groups being superimposed upon the corresponding vibrations of CH_3 groups at the resolution achieved by the usual type of routine instrument.

- - - - - - - - - - - - - - - - - - - -

A270 (a). The absence of absorption just above 3000 cm^{-1} rules out (b) and (c). (There are other, more reliable, ways of detecting alkenic and benzenoid systems but these will be discussed later.)

F272 Does the spectrum allow us to eliminate cyclic ketones?

- - - - - - - - - - - - - - - - - - - -

A271 (c). This point was dealt with in Part 1 (**F69**).

The higher frequency component is almost invariably the stronger which supports its assignment to the antisymmetrical vibration.

F273 What types of cyclic ketone are excluded?

- - - - - - - - - - - - - - - - - - - -

A272 No, not entirely.

F274 The compound is thus probably a simple aliphatic ketone though it may contain a medium or large ring carrying one or more side-chains. Knowledge of its molecular formula—or even just its molecular weight—would probably enable us to complete the identification.

Given that the molecular formula is C_4H_8O identify the compound.

A273 4- and 5-ring ketones. The unusual strength of the band at 1380 cm^{-1}, presumably due to the symmetrical CH_3 deformation, compared with the other CH deformations rules out all simple ring ketones which contain only CH_2 groups. It also suggests that the alkyl groups are highly branched or else small.

A274 The compound is 2-butanone, $CH_3COCH_2CH_3$.

Problem 2

Identification of a colourless liquid C_7H_9N

F275 It is convenient at this point to note the information that can be deduced from the molecular formula alone. We can determine simply by inspection the number of 'double bond equivalents' in the molecule. By double bond equivalent we mean any feature which affects the molecular formula in the same way as the presence of a double bond. Thus the following are included:

(i) double bonds themselves (>C=C<, >C=N—, >C=O)

(ii) rings, e.g. [cyclopentane], [tetrahydropyran] etc.

(iii) triple bonds which count as 2 double-bond equivalents

(—C≡C—, —C≡N)

The number of double bond equivalents (DBE) may be readily obtained using the formula

$$DBE = 1 + N_C - \tfrac{1}{2}N_H - \tfrac{1}{2}N_X + \tfrac{1}{2}N_N$$

where N_C, N_H, N_X and N_N are the numbers of carbon, hydrogen, halogen and nitrogen (trivalent) atoms respectively.

Note that no account is taken of oxygen. The calculation cannot be applied to compounds containing sulphur because of that element's variable valency. N—O compounds can also cause trouble occasionally.

Write down the number of double bond equivalents in the following molecules:

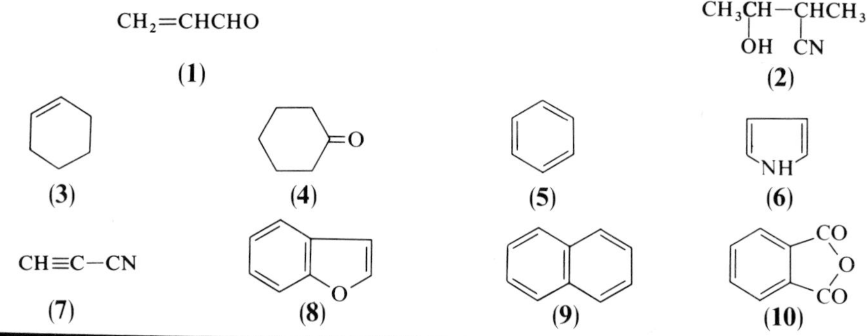

F276 Write down the number of double bond equivalents in the following molecules: (a) $C_5H_{10}O$, (b) $C_7H_{13}NO$, (c) C_8H_9N, (d) C_7H_8NCl, (e) C_4H_3BrO, (f) $C_2H_3NO_2$.

- -

A274 (**1**) 2, (**2**) 2, (**3**) 2, (**4**) 2, (**5**) 4, (**6**) 3, (**7**) 4, (**8**) 6, (**9**) 7, (**10**) 7.

F277 The number of double bond equivalents obtained from the formula can usually be satisfied by several different possible structures. The infrared spectrum however will enable these to be quickly checked.

What combinations of double bond equivalents could be present in formula (b) of **F276**?

- -

A276 (a) 1 (b) 2 (c) 5 (d) 4 (e) 3 (f) 2

F278 Since a benzenoid ring corresponds to 4 double bond equivalents a high DBE value will immediately suggest that the molecule is aromatic. Which of the formulae given in **F276** could arise from aromatic compounds?

- -

A277 (i) Two double bonds: both C=C, C=C and C=O, C=C and C=N or C=N and C=O.

(ii) Two rings. The hetero atoms could be incorporated in either ring, both rings or neither ring, e.g.

(iii) One ring and one double bond. Again the hetero atoms could be present in the ring or the double bond, e.g.

The double bond could also be endocyclic (within the ring) e.g.

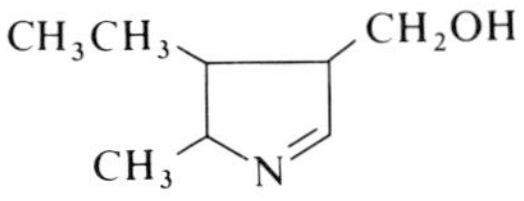

$$\begin{array}{c} CH_3CH_3 \diagdown \quad \diagup CH_2OH \\ \text{(ring)} \\ CH_3 \diagup \; N \end{array}$$

(iv) A triple bond C≡C or C≡N.

F279 Now let us return to the present problem. The compound has the molecular formula C_7H_9N. Is it possible for this compound to possess a benzenoid ring?

- -

A278 (c) and (d). More often than not a DBE value of 4 or more will indicate that the compound is benzenoid. Note also the other requirement that the compound must have six or more carbon atoms.

F280 In most cases the important question of the presence or absence of a benzenoid ring is readily settled by reference to the infrared spectrum.

Write down the regions of the spectrum in which you would expect to find absorption if the compound is benzenoid.

Which of these is **least** useful for characterization purposes?

- -

A279 Yes. The molecule has 4 double bond equivalents

F281 The most conclusive evidence for aromaticity is found in the 1430—1630 cm^{-1} region. These bands are due to ring stretching and for the majority of organic compounds two bands are observed one near 1600 cm^{-1} and the other near 1500 cm^{-1}. Although alkenic C=C stretching occurs in roughly the same frequency range these bands are usually relatively weak compared with the aromatic bands and only exceptionally occur as low as 1500 cm^{-1}. The spectrum of the compound C_7H_9N is shown in Fig. 5.2.

Write down the frequencies of bands in the 1430—1630 cm^{-1} region in the spectrum which could be assigned to a benzenoid ring.

- -

A280 3000—3100; 1430—1610; 1000—1280; 570—900 cm^{-1}. The in-plane deformations (1000—1280 cm^{-1}) are least useful since they occur in the skeletal region.

F282 If the compound is aromatic we should observe medium to strong absorptions in the C—H out-of-plane bending region. Write down the frequencies of any bands which you consider could be assigned to out-of-plane C—H bending modes.

- -

A281 1480, 1520, 1605 cm^{-1}.

F283 Consider now the C—H stretching region. Which of the bands in this region could be due to benzenoid C—H stretching?

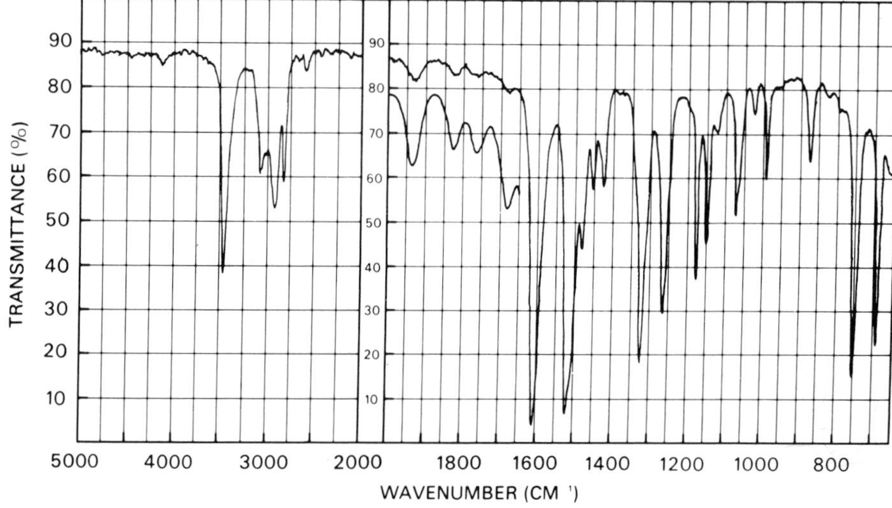

Fig. 5.2. (F281).

A282 695 and 750 cm^{-1}, both strong bands.

F284 Returning to the molecular formula at this point, we note that the molecule contains a nitrogen atom. Write down the different kinds of group in which the single nitrogen atom could be incorporated into this molecule.

A283 The band at 3060 cm^{-1}. Note the weakness of this band compared with other benzenoid bands.

F285 If the compound contains a benzenoid ring, which of these groups are eliminated by the molecular formula?

A284 $-NH_2$ $>NH$ $>N-$ (tertiary and heterocyclic)
$-CN$ and $-NC$, $>C{=}N-$.

Note that the absence of oxygen rules out amide ($-CONH_2$) and groups such as nitro, ($-NO_2$) and cyanate ($-CNO$).

F286 Which of the remaining groups will give rise to reliable characteristic bands. Indicate the region in which they will be found.

A285 $-C{\equiv}N$, $-N{\equiv}C$ and $>C{=}N$. These require DBE values of 2, 2 and 1 respectively and we have 'used up' the 4 double bond equivalents present with the benzenoid ring.

F287 Are either of these possibilities supported by the spectrum?

A286 NH_2 and NH, 3180—3500 cm^{-1}.

F288 What does the presence of absorption just below 3000 cm^{-1} indicate?

A287 Yes. The strong sharp band at 3450 cm^{-1} confirms the presence of $>$NH or $-NH_2$. The frequency of the NH stretching band does not enable us to distinguish between primary and secondary amines but an experienced spectroscopist would be able to decide at once between these two possibilities from the general shape of the band. We shall, however, defer discussion of this point until later.

F289 The bands at 2800 and 2900 cm^{-1} comprise the usual doublet which is observed for 'aliphatic C—H'. The splitting is in fact rather more pronounced than usual.

Thus we have established the presence of the following structural features in the molecule:

aliphatic C—H, $\rangle$N—H, benzenoid ring

Write down all the possible structures for the molecule consistent with the molecular formula and containing all these features.

A288 The molecule contains aliphatic C—H groups as well as aromatic C—H.

F290 One way of distinguishing between these possibilities is to determine the substitution on the benzenoid ring. In which region of the spectrum shall we seek evidence on this point?

A289 The molecular formula and the presence of the benzenoid ring leave only one carbon atom to be accounted for. This gives us the following possibilities:

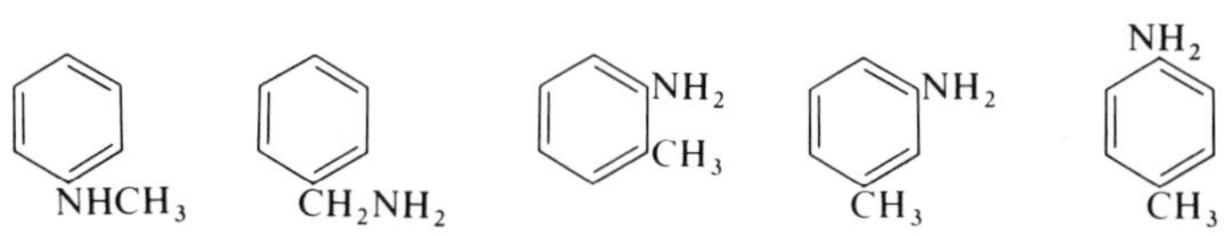

$NHCH_3$ CH_2NH_2 NH_2 / CH_3 NH_2 / CH_3 NH_2 / CH_3

F291 Which of the following types of substitution give the closest fit to the observed spectrum:

mono, 1,2-di-, 1,3-di-, 1,4-di-?

A290 The C—H out-of-plane bending region 670—900 cm^{-1}.

F292 Further support for monosubstitution is provided by the appearance of four bands and even spacing in the 1650—2000 cm^{-1} region. This pattern, which is well defined in such molecules as toluene and polystyrene, is easily recognizable if the general intensity is high enough. (In Fig. 5.2 this region has been run twice, once using a much thicker cell than was used for the rest of the spectrum.) Other types of substitution also give characteristic patterns in this region which are sometimes regarded as more reliable than the C—H out-of-plane deformations.

Which of the following statements regarding the origin of these bands is true?

(a) They are overtones of the out-of-plane C—H deformations.
(b) They are combinations involving out-of-plane C—H deformations.

A291 mono. The appearance of the 695/750 cm^{-1} bands is very characteristic of monosubstitution.

1,3-disubstitution is also possible though the band at 875 cm^{-1} is rather weak for this interpretation. Its presence does not conflict with monosubstitution because it might then be due to a vibration of the side-chain.

1,2-disubstitution is ruled out because, although two bands occur in the right region, the lower frequency band, 695 cm^{-1}, is not a weak absorption.

1,4-disubstitution is very unlikely since the 875 cm^{-1} band is weak and of rather high frequency. The presence of the two strong bands below 800 cm^{-1} also conflicts with the expected pattern for 1,4-disubstitution.

F293 If the molecule does contain a monosubstituted benzenoid ring (which seems most likely from the evidence) we are left with the choice between the primary amine

C_6H_5—CH_2NH_2

and the secondary amine

C_6H_5—$NHCH_3$

This distinction is easily made since the primary amine group, —NH_2, gives rise to a broad NH stretching band which is frequently a clearly resolved doublet.

The corresponding band for the secondary amine group >NH however is sharp and single. Which of the two possibilities is therefore correct in the present case? Write down the formula of the compound.

A292 (b). Overtones would appear close to twice the frequency, i.e. 1500—1600 cm^{-1}.

F294 How do you account for the fact that the N—H stretching band of primary amines is a doublet?

A293 The compound is a secondary amine, *N*-methylaniline

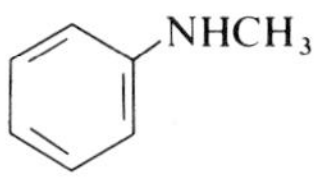

$C_6H_5NHCH_3$

The unusual splitting of the C—H stretches referred to in **F289** also supports this conclusion. Incidentally, low CH frequencies (2810) are typical of N—CH_3.

F295 Like the CH_2 group, the NH_2 also has a bending frequency ('scissoring') occurring in the range 1490—1630 cm^{-1} which is often useful for the recognition of the group.

Why is this of no assistance to us in the present example?

A294 The NH_2 group like the CH_2 group has two stretching vibrations, one symmetrical the other antisymmetrical.

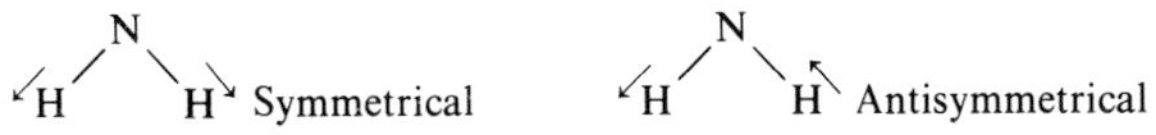

N(H)(H) Symmetrical   N(H)(H) Antisymmetrical

The greater breadth of the NH_2 band is accounted for by the greater possibility of hydrogen bonding with the NH_2 group than with >NH.

A295 The region contains strong bands due to the aromatic ring stretching. These would obscure the NH_2 band which frequently occurs near 1600 cm^{-1}. The deformation of the NH bond in R_2NH is ill-defined and does not give rise to a useful band.

Problem 3

Identification of a white solid m.p. 54°C, M. Wt. 80

F296 In this example we have no knowledge of the molecular formula. We do not even know whether the compound is organic or inorganic. What features of the spectrum (Fig. 5.3) confirm that we are in fact dealing with an organic compound?

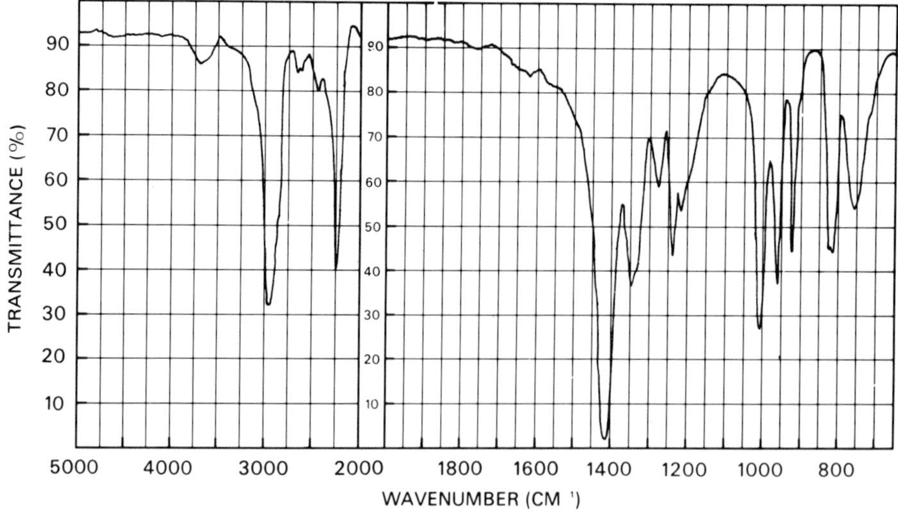

Fig. 5.3

F297 What information can we obtain about functional groups from bands above 1500 cm^{-1}?

A296 The compound is organic (or contains an organic part) since bands due to CH stretching and bending appear with high intensity. (The low melting point also suggests that the compound is more likely to be organic than inorganic).

F298 The position of the band does not enable us to choose between the possibilities although we note that 2250 cm^{-1} is towards the upper limit for C≡C. Let us stay with C≡C for a moment and see whether or not we can eliminate it from the list in the light of the i.r. spectrum. If the molecule is an alkyne the triple bond must be present as R—C≡CH or R—C≡C—R. Which of these is eliminated by the spectrum?

A297 A strong sharp band at 2250 cm^{-1} falls in the ranges of CD, C≡C, C≡N and NCO. (We have already noted the presence of CH). Of equal importance, the **absence** of bands characteristic of OH, NH and C=O eliminate these groups. (The weak feature at 3700 cm^{-1} is too broad to be non-bonded OH.)

F299 Of the four groups in **A297** would you expect —C≡C— to give rise to

(a) the strongest band
(b) the weakest band
(c) a band of intermediate intensity?

A298 RC≡CH. There is no band near 3300 cm^{-1} which could be ascribed to the C—H stretching vibration.

F300 In the light of the spectral evidence does the compound contain a C≡C bond?

A299 (b). The band will be particularly weak in disubstituted alkynes. If the molecule were symmetrical it would be completely absent. The spectra of 1-hexyne and 2-hexyne are shown in Fig. 5.4.

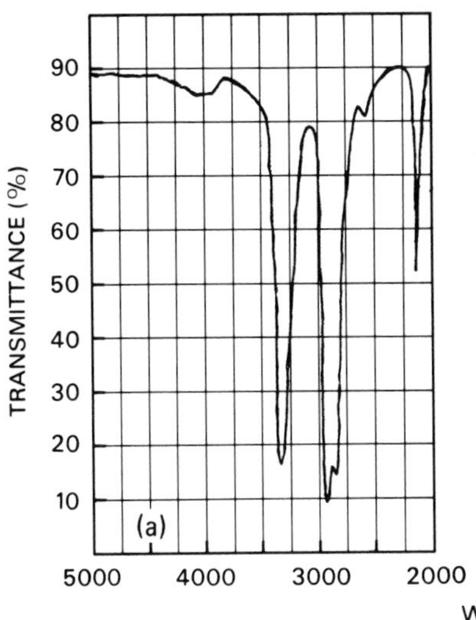

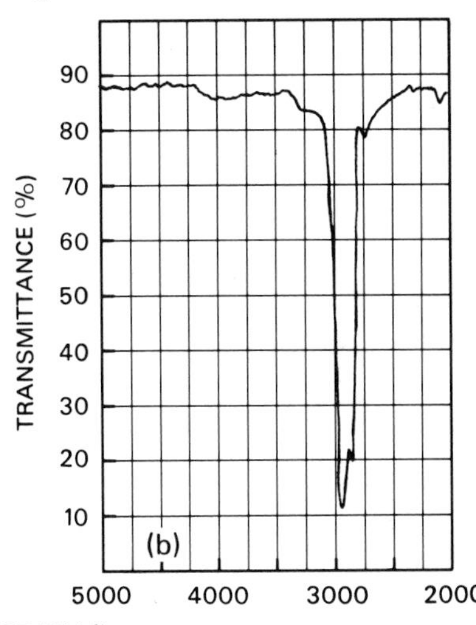

WAVENUMBER (CM^{-1})

Fig. 5.4. Partial spectra of **(a)** 1-hexyne; **(b)** 2-hexyne.

F301 Bearing in mind the restrictions imposed by the molecular weight let us turn to the —NCO possibility. What is the largest alkyl group that could be associated with this group?

A300 No, the 2250 cm^{-1} band is relatively strong as well as being of rather high frequency for RC≡CR. Incidentally, the physical properties of the compound also support this view since an alkyne of M. Wt. 80 would not be a solid unless it contained a polar group such as —COOH or —$CONH_2$ and this is not consistent with the spectrum.

The nearest acetylene to this molecular weight is $C_3H_7C{\equiv}CCH_3$(2-hexyne) (M. Wt. 82) which is a liquid, b.p. 92°C.

F302 Quite apart from the difficulty over the molecular weight there is strong spectral evidence for rejecting —NCO.

A partial spectrum of isopropyl isocyanate is shown in Fig. 5.5. What do you think this strong spectral evidence is?

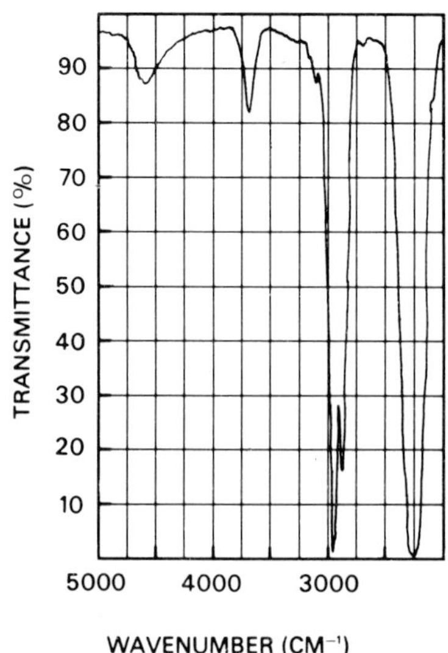

Fig. 5.5. Partial spectrum of isopropyl isocyanate.

A301 C_2H_5, ethyl. $C_3H_7NCO = 85$ is in fact closer to the molecular weight but the discrepancy suggests that the compound is not a simple isocyanate.

F303 We now have to choose between —CN and CD. Although we cannot come to any definite decision about CD from spectral evidence except to say that the band lies too high in the range to be due to aliphatic CD (cf. the distribution of CH frequencies) we need not consider this possibility further. Compounds containing deuterium at a level high enough to make its presence apparent in an infrared spectrum do not occur naturally nor in materials of industrial origin. (The natural abundance of the ^{2}H isotope is 0.015% of the common isotope ^{1}H). A spectroscopist encountering a deuterated compound will almost invariably have had some indication that deuterium may be present from the origin of the sample. Under normal circumstances he would not regard the presence of C—D as a serious possibility. Continue to **F304**.

A302 The isocyanate band in a molecule of comparable size is very much stronger than the C—H bands; it is in fact the strongest band in the spectrum. The —NCO band is also very broad (in contrast to CN bands). This illustrates the importance of taking other properties of the band than its frequency into account when using it for diagnostic purposes. So NCO is not present.

F304 Having eliminated NCO, is there any other way, consistent with the spectrum, that oxygen could be present in the molecule?

F305 We have thus established the presence of CH and CN in the molecule and also the absence (equally important) of OH, NH and CO. Is the molecular weight consistent with the compound being a simple aliphatic cyanide?

(For the purposes of this problem you may assume that the molecular weight is 80 ± 0.5 (i.e. accurate to 2 figures).)

- - -

A304 No. There is no evidence of any of the common functional groups containing oxygen such as OH, CO and C—O nor for the less common groups such as N=O and S=O which give rise to very strong infrared bands.

F306 Although we could get nearer the molecular weight by introducing one or two double bonds or rings (or one triple bond) into the molecule none of the possibilities reproduces the value exactly (the molecular weight of these unsaturated compounds, like the saturated compound, must be an odd number).

The fact that the molecule has an even molecular weight is highly significant since we already know that it contains a nitrogen atom. Furthermore, all compounds containing only C, H and one N atom are of odd molecular weight so our molecule must contain another hetero atom. Which of the following groups represented by X in the formula $\mathrm{R}\underset{\mathrm{X}}{\underset{|}{\mathrm{C}}}\mathrm{HCN}$ where R = H or alkyl could give rise to an even molecular weight for the compound?

(a) —Cl (mass 35.5)
(b) —F (mass 19)
(c) A group containing S (mass 32)
(d) A group containing N (mass 14).

- - -

A305 No. C_4H_9CN (M.Wt. 83) is closest.

F307 Notice that chlorine and fluorine cannot be excluded on spectral evidence since bands occur in the range of both C—Cl and C—F.

Only two groups containing nitrogen are in fact possible in view of the infrared spectrum. What are they?

- - -

A306 (d). Cl and F would lead to an odd molecular weight because the sum of the masses of all the other atoms must be even (R must be odd;

CHN = 27). A group containing S must have an odd mass because S itself has even mass and is di-, tetra- or hexavalent. In fact it turns out that an organic compound of **even** molecular weight containing nitrogen must have an **even** number of nitrogen atoms. This rule holds even if other hetero atoms are present. Unlike N these will either be (a) even atomic weight, even valency (S, O, Si) or (b) odd atomic weight, odd valency (F, Cl, Br, I, P). —SR, $-SiR_3$, $-PR_2$ and halogen atoms all have odd mass whereas X must have even mass if the molecular weight is to be even since RCHCN (with a bond from CH) has even mass.

F308 The simplest tertiary amine structure

$$(CH_3)_2N-CH_2CN$$

has a molecular weight 84. How could this be reduced to the correct value of 80?

A307 Since NH is absent, tertiary amine and a second C≡N group are the only possibilities.

F309 Draw out the structures of possible unsaturated cyclic formulas of the correct molecular weight. How can they be eliminated by the spectrum?

A308 Introduce two double bonds or one double bond and one ring. (Two rings is unlikely, whilst HC≡C— has already been eliminated).

F310 Two compounds containing two CN groups and having a molecular weight of 80 are possible. What are they and how could we decide which one we have using i.r. spectroscopy?

A309

CH_2-CHCN / $N{=}CH$ (ring)	$CH-CHCN$ / $CH-NH$ (ring, CH=CH)	$CH-CH_2$ / $CH-NCN$ (ring, CH=CH)
No C=N band	No NH band	No C=C band

A310

CH_2CN–CH_2CN Succindinitrile

$CH_3CH(CN)_2$ 1,1-Dicyanoethane

One could argue that the absence of a band at 1380 cm^{-1} (CH_3) indicates that the compound cannot be 1,1-dicyanoethane. However the two possibilities cannot be distinguished with certainty by considering the spectrum alone, but it is a very straightforward matter to compare this spectrum with reference spectra of the two compounds to identify which one we have. In fact the compound whose spectrum is given is succindinitrile (1,2-dicyanoethane).

Notice that we were able to narrow down the choice to two compounds from the i.r. spectrum without knowledge of the molecular formula. Knowledge of the molecular weight however was a key factor giving us an idea of the size of the molecule and an indication of the relative intensities of group frequencies which might be expected. It also enabled us to exclude certain classes of compounds from our general knowledge of the physical properties of these compounds given that the unknown was a low melting solid. To be of maximum use the molecular weight must be accurate to at least one mass unit. The only method for determining the molecular weight to this accuracy is mass spectroscopy. This technique also provides much information about the molecular structure and is frequently used alongside the infrared spectrum together with other data in structural diagnoses.

Problem 4

Identification of a colourless liquid b.p. 195°C, molecular formula $C_7H_{16}O$

F311 This example, Fig. 5.6, is very straightforward and you should quickly be able to establish what class of compound we are dealing with. What functional group contains the oxygen atom?

F312 Is the compound aromatic or aliphatic?

- - - - - - - - - - - - - - - - - - - -

A311 OH, characteristic band 3360 cm^{-1}.

F313 The compound is thus a saturated alcohol. We now have to try to settle whether it is primary, secondary or tertiary, and whether the carbon chain is straight or branched. The former task can be carried out by considering the frequency of the C—O band. Locate this band and write down its frequency.

- - - - - - - - - - - - - - - - - - - -

A312 Aliphatic—no strong bands below 900 cm^{-1};
no bands near 1500—1600 cm^{-1};
no aromatic C—H stretching above 3000 cm^{-1}.

But note that we do not need the spectrum to establish this fact since the formula indicates that the compound must be open chain and saturated (C_nH_{2n+2}).

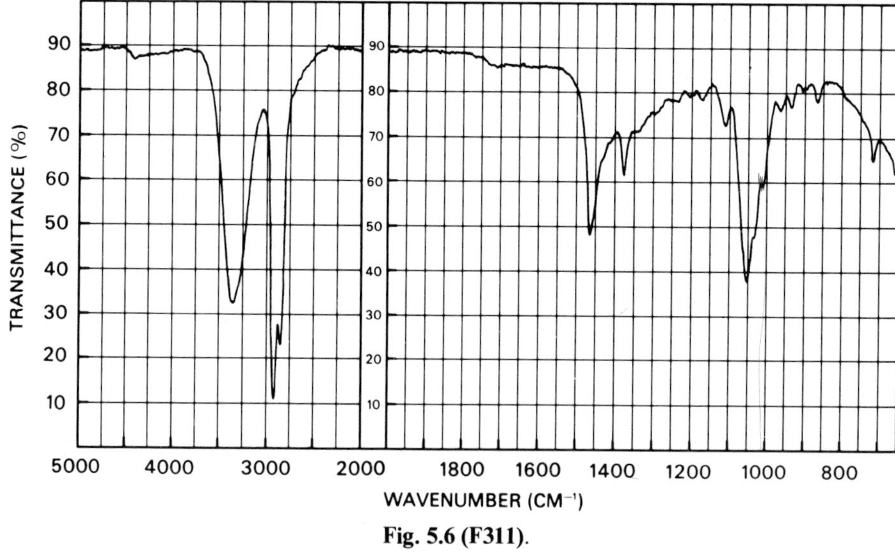

Fig. 5.6 (F311).

F314 We have frequently referred to the fact that the position of bands in the skeletal region is very much influenced by the overall structure of the molecule. As a result one usually identifies the C—O band by its intensity rather than its frequency (which is generally of no diagnostic value). However, useful structure-frequency correlations have been established for saturated alcohols which enable one to differentiate between primary, secondary and tertiary alcohols. The ranges concerned are given in Chart 5 (A). Refer to this and decide which type of alcohol we are dealing with.

A313 1050 cm^{-1}. (Note the high intensity relative to C—H bending at 1460 cm^{-1}).

F315 The correlations in Chart 5 (A) must be used with caution. Possible complications due to the presence of other functional groups must always be borne in mind.

In which of the following compounds would identification of the alcohol as primary, secondary or tertiary be unreliable or ambiguous?

$CH_3CH{=}CHCH_2CH_2CH_2OH$

(11)

$CH_3CH_2C(OH)(CH_3)-CH_2CH_3$

(12)

$CH_3CH_2OCH_2CH_2CH_2OH$

(13)

$CH_3CH(Cl)CH_2CH_2CH(CH_3)OH$

(14)

A314 Primary alcohol.

F316 We are now left with the problem of determining whether the carbon chain is straight or branched (containing side-chain methyl or larger groups). Infrared spectroscopy is not well-suited for this purpose, and it is usually better to turn to other techniques such as n.m.r. spectroscopy. There are however two correlations given in Chart 5 (B) which identify certain features of the alkyl group. Are either of these features present on the evidence of Fig. 5.6?

A315 (**13**) has a second C—O group which will also produce a strong band in the 1000—1300 cm^{-1} region. It will not be possible therefore to identify with certainty which is the C—OH band.

In (**11**) the C=C group is too far from the C—OH group to affect its frequency.

F317 Write down as many structures as you can for a primary alcohol containing seven carbon atoms, no *gem*-dimethyl structure and four or more adjacent CH_2 groups.

A316 Yes. There is a band at 720 cm^{-1} indicative of four or more adjacent CH_2 groups. Note the low intensity of this band which renders it easily obscured by other bands in this region—notably aromatic C—H out-of-plane bends.

The 1380 cm^{-1} band is clearly single. The characteristic doublet confirming *gem*-dimethyl is illustrated in Fig. 5.7 where it is due to the isopropyl group in isopropanol.

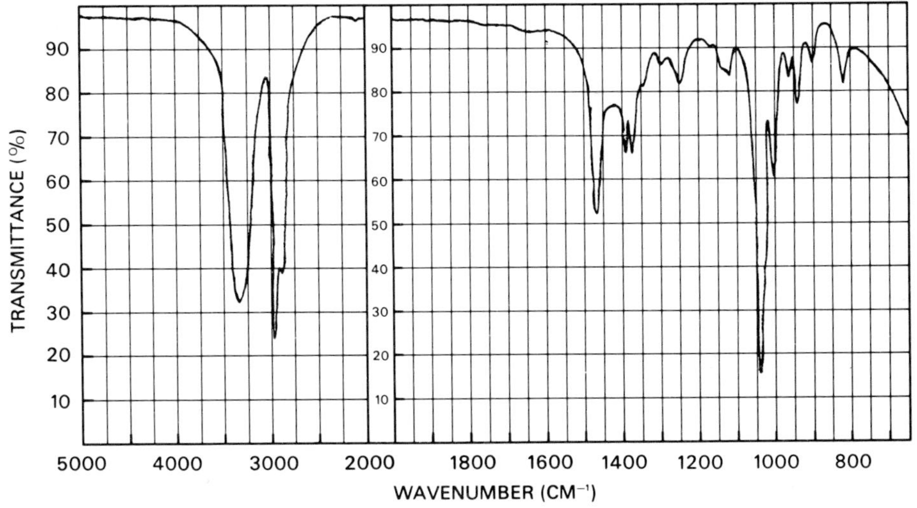

Fig. 5.7

A317 $CH_3CH_2CH_2CH_2CH_2CH_2CH_2OH$, 1-heptanol is the only possibility.

5-methyl-1-hexanol has an isopropyl group whilst 4-methyl-1-hexanol and 3-ethyl-1-pentanol have less than 4 adjacent CH_2 groups.

It must be emphasised that in practice it is very unusual to identify a compound using the infrared spectrum alone. Normally it is necessary to look to some other spectroscopic technique such as nuclear magnetic resonance or mass spectroscopy to settle details of the skeleton of a compound. The present example is atypical in that the infrared spectrum alone was sufficient to make a positive identification.

Problem 5

Identification of a liquid isolated from the chlorination product of a mixture of hydrocarbons. M. Wt. 273; carbon content 22%

F318 What is unusual about the appearance of the spectrum (Fig. 5.8) compared with the majority of spectra of organic compounds?

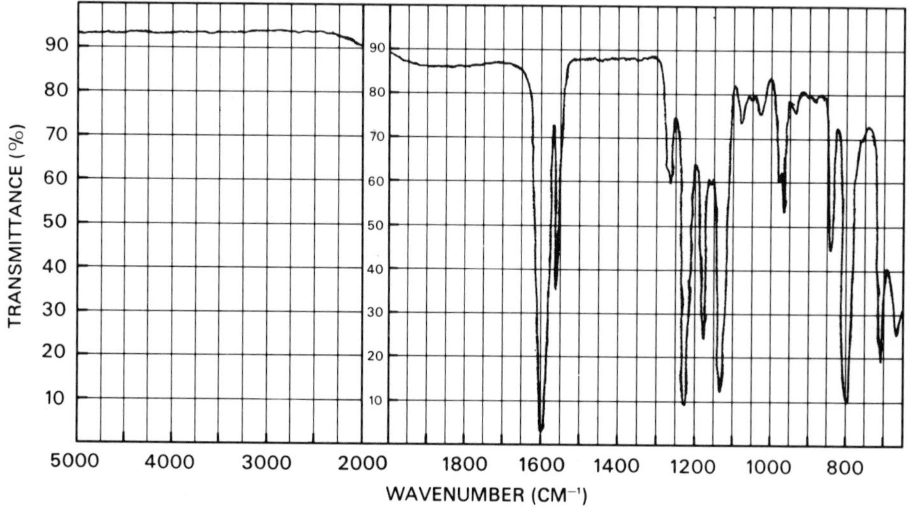

Fig. 5.8

What conclusion do you draw from this?

F319 You should be able to assign two of the principal bands with confidence, bearing in mind the absence of hydrogen and the origin of the compound.

A318 There is no absorption above 1700 cm^{-1} nor between 1300 and 1500 cm^{-1}. Hence there is no hydrogen in the molecule.

F320 Since the compound contains 22% carbon, 22% of the molecular weight (60 atomic mass units) must be due to carbon atoms, five in number. We thus have a C_5 molecule containing chlorine, no hydrogen and at least one double bond. What is the maximum number of chlorine atoms we can incorporate in the molecule consistent with the molecular weight.

- - -

A319 1600 cm^{-1} C=C stretching;
800 cm^{-1} C—Cl stretching.

The latter is quite unambiguous in this case since the bands which normally interfere with it are associated with various types of CH deformation. These of course cannot occur in this molecule.

F321 Since the formula C_5Cl_6 gives us precisely the correct molecular weight it seems reasonable to accept this as the most likely possibility. With the possible exception of oxygen, all other atoms are excluded by the method of preparation. It would not be possible to reproduce the molecular weight as exactly if we incorporate oxygen into the molecule.

Assuming the formula to be C_5Cl_6 calculate the number of double bond equivalents in the molecule.

What is the most likely combination of double bond equivalents consistent with the spectrum?

- - -

A320 Six. $\underset{(5 \times 12)}{273 - 60} = 213 \qquad \frac{213}{35.5} = 6$

F322 Write down possible structures for the molecule on this basis.

- - -

A321 Three. C≡C and cumulated C=C are unlikely in view of the total absence of bands above 1700 cm^{-1} and so the molecule cannot be open chain. This leaves the combination of two double bonds and one ring, or two rings and one double bond. Of these the latter seems less likely in view of the small number of carbon atoms.

F323 Routine application of infrared spectroscopy cannot distinguish unambiguously between these structures, particularly in view of polychlorination, the effect of which on C=C stretching frequencies in strained cyclic alkenes is not easy to predict. An experienced chemist would recognize the implausibility of some of the structures but strictly speaking this is as far as infrared spectroscopy could go with the problem. The compound is in fact hexachlorocyclopentadiene. In the light of this, how do you account for the three strong bands between 1100 and 1250 cm^{-1}?

- - -

A322

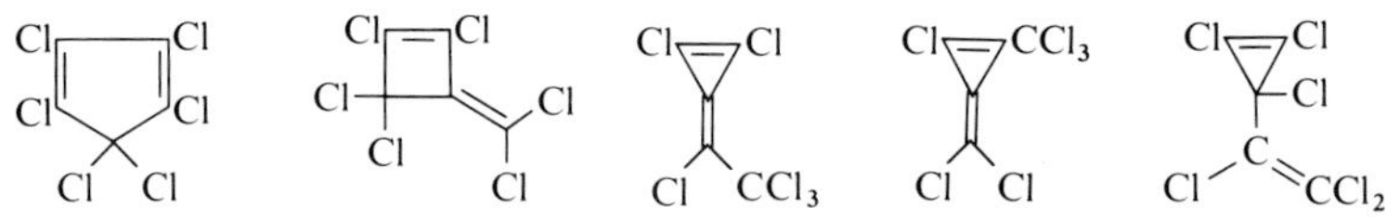

F324 How do you account for the relatively high intensities of these normally weak bands?

A323 They must be due to C—C stretching since C—Cl stretches and all bending vibrations have frequencies below 900 cm^{-1}.

Three C—C stretching bands are to be expected since there are 3 C—C bonds in the molecule.

C—H rocking bands which normally obscure this region of the spectrum are absent in this case and, most unusually, this allows us to identify C—C stretches. (The weak band at 1270 is probably an overtone or combination.)

A324 The presence of $>CCl_2$ at one end of the molecule makes the bonds more polar than in hydrocarbon fragments.

Problem 6

Identification of a colourless liquid b.p. 132 °C; M. Wt. 112

F325 Decide whether the molecule whose spectrum is given in Fig. 5.9. is

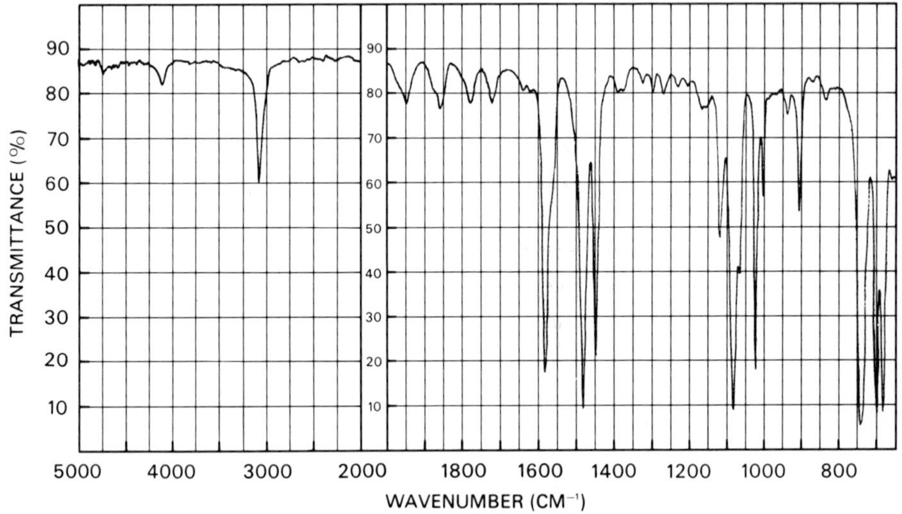

Fig. 5.9

(a) entirely aliphatic
or (b) aromatic but containing alkyl groups
or (c) entirely aromatic.

F326 What conclusion do you draw about the substitution of the benzenoid ring? Which of the following applies?

(a) monosubstituted
(b) *o*-disubstituted
(c) *m*-disubstituted
(d) *p*-disubstituted
(e) not clear from the spectrum.

A325 (c). There are no bands due to aliphatic C—H bands.
Note the aromatic CH stretching band, ring stretching frequencies near 1500 and 1600 cm^{-1} and CH out of plane deformations near 700 cm^{-1}.

F327 What can you say about the nature of the substituent?

A326 (a). Note the strong bands at 690 (double) and 740 cm^{-1}. The regular pattern of bands between 1700 and 2000 cm^{-1} is also very characteristic of monosubstitution as we pointed out earlier (**F284**). These bands are more prominent here because the spectrum has been run using a cell of path length 0.02 mm instead of the customary capillary film.

F328 Now use the molecular weight to identify the compound.

A327 OH, NH_2 and aliphatic CH are all eliminated by the absence of bands in the high frequency region. The absence of carbonyl absorption eliminates —CHO, COOH, COCl etc.

No positive indication.

A328 The phenyl group accounts for 77 mass units. The remaining 35.5 points directly to chlorine. Since other substituents which could conceivably make up the molecular weight have been eliminated the compound is identified as chlorobenzene. Notice that we have not identified a band due to C—Cl in the spectrum since this would fall in the same region as the strong out-of-plane CH deformations of the aromatic ring. It is possible however that the unusual splitting of the lower frequency band in this region can be attributed to the presence of C—Cl.

Problem 7
Identification of a colourless liquid M. Wt. 86

F329 Identify the compound whose spectrum is given in Fig. 5.10.

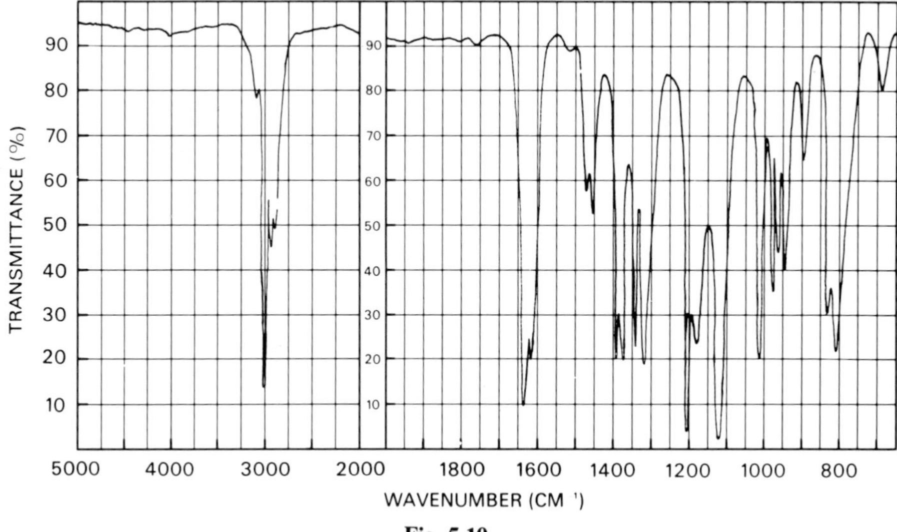

Fig. 5.10

F330 Having put forward a possible identification for the unknown compound it is wise to look at the spectrum again to see if any more features can be tied in with the proposed structure. In particular we should look for bands associated with the vinyl group. What shall we expect to find? (Refer to Chart 3).

A329 C—H region indicates presence of both aliphatic and aromatic or alkenic C—H groups. Aromatic compounds however are excluded by the low molecular weight.

The band at 1640 cm^{-1} is probably due to C=C in view of the evidence of the 3050 cm^{-1} band for alkenic CH. The only other possible interpretations of this band would be as some low frequency carbonyl stretch, for example, an amide.

(This could not be primary or secondary since there is no N—H stretching. The tertiary amine CH_3CONMe_2 (M. Wt. 87) is close to the given molecular weight but we are then left with the alkenic CH band to explain.)

Accepting that 1640 cm^{-1} is a C=C band, its unusually high intensity indicates that the bond must be polar and probably has an electronegative group attached to it.

The strong band at 1120 cm^{-1} points to the presence of C—O which in the absence of evidence for OH and C=O (ester) can only be an ether link C—O—C. This might well account for the polarity of the C=C bond.

The strong doublet at 820 cm^{-1} is probably an out-of-plane bend of the alkenic C—H which we know to be present. Although the frequency falls in the range of C—Cl the low molecular weight and the total mass of the fragments already identified exclude the presence of chlorine.

We have now accounted for all the characteristic bands (including 1120 cm^{-1}) but we have not yet exhausted all the evidence available from the spectrum. The splitting of the 1380 cm^{-1} band as we saw in an earlier problem is indicative of the presence of isopropyl or some other structure containing a *gem*-dimethyl group. Thus we have established the presence of the following groups:

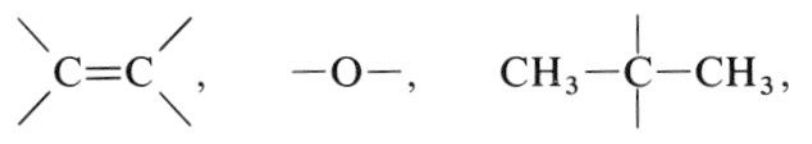

$$>C=C<, \quad -O-, \quad CH_3-\overset{|}{\underset{|}{C}}-CH_3,$$

a total mass of 82. The remaining four mass units can only be made up with hydrogen atoms which nicely satisfy the remaining positions in the structure obtained by joining the three groups together. The compound is therefore isopropyl vinyl ether.

$$CH_2=CH-O-CH(CH_3)_2$$

F331 Locate these bands in the spectrum in Fig. 5.10.

- - - - - - - - - - - - - - - - - - - -

A330 Two bands between 900 and 1000 cm^{-1} and one between 600 and 700 cm^{-1} due to out-of-plane CH vibrations.

(=CH and C=C stretching and in-plane C=CH bending vibrations are not specific to the vinyl group.)

F332 How do you reconcile the appearance of the bands in the CH out-of-plane bending region compared with what is expected for the vinyl group?

- - - - - - - - - - - - - - - - - - - -

A331 Four weak bands occur between 900 and 1000 cm^{-1} and a very weak band at ~700 cm^{-1}. The most prominent feature in this region is the doublet near 820 cm^{-1} which is well outside all the vinyl ranges.

F333 Suggest a reason for the splitting of the C=C stretching band 1620 + 1640.

- - - - - - - - - - - - - - - - - - - -

A332 Detailed assignments, particularly of CH out-of-plane deformations which are generally the most useful, cease to be reliable when the double bond is adjacent to a polar group.

A333 Fermi resonance with the overtone of the band at 820 cm^{-1}.

In all the examples you have met so far sufficient information has been given in addition to the spectrum to enable a positive identification of the compound to be made. This situation is not as unrealistic as might be supposed in practice with the growing availability of mass spectrometers, improved apparatus for determining molecular weights and automatic carbon/hydrogen analysers, not to mention other spectroscopic methods such as nuclear magnetic resonance. More frequently however the chemist, particularly the chemist working in a small industrial laboratory, does not have immediate access to all these resources. Even if he does he is frequently called upon to identify a sample quickly and cannot wait for the results of three or four different analytical methods to become available. In these circumstances infrared spectroscopy is a particularly useful method because the spectrum can be obtained quickly and easily and yields a great deal of information. It is for this reason that the technique has become so widespread, an infrared spectrometer being frequently the first (and in some cases the only) analytical instrument to be acquired by an industrial laboratory.

The infrared spectrum will usually enable the chemist to determine the class of compound to which any unknown sample brought to him belongs. This is frequently quite sufficient to enable him to narrow down the possibilities for its structure to a small number, particularly if he has specialist knowledge of the origin of the sample or, in the case of a commercial product, the purpose for which it is intended (e.g. plasticizer, solvent, detergent, insecticide, lubricating oil etc.). Comparison of the spectrum of the sample with a small number of spectra of reference compounds (collections of which may be purchased or built up by the infrared laboratory over the years) will usually result in the identification of the compound.

We conclude this programme with some problems in which you are required only to establish the class of compound concerned.

Problem 8

F334 The spectrum shown in Fig. 5.11 is that of a waxy solid. What conclusions can you draw regarding its nature?

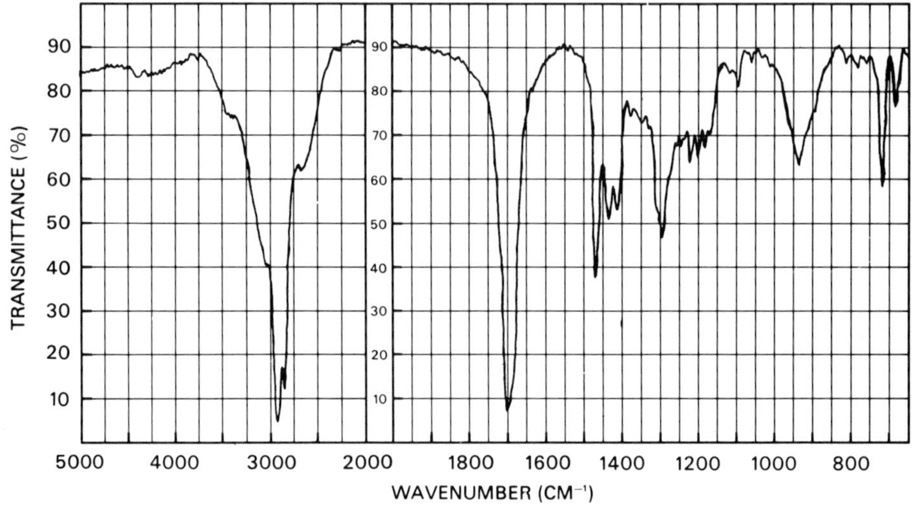

Fig. 5.11

A334 The compound is aliphatic (no aromatic bands) and probably saturated. (There are no obvious bands characteristic of unsaturated entities though one double bond in a long-chain hydrocarbon fragment may not be sufficient to provide infrared evidence of unsaturation).

It is a carbonyl compound, the carbonyl band occurring in the range characteristic of saturated carboxylic acids. The appearance of the spectrum in the region 2400—3600 is also typical of COOH. The band at 720 cm^{-1} indicates the presence of a straight chain of four or more CH_2 groups. (The splitting apparent in this band is due to the sample being run as a solid. In the liquid phase or solution a single band is obtained).

The compound is therefore a long chain fatty acid which would in fact be identified as stearic acid, $CH_3(CH_2)_{16}COOH$, by comparison with reference spectra of this class of compound.

Problem 9

F335 To which of the following classes does the polymer whose spectrum is given in Fig. 5.12 belong?

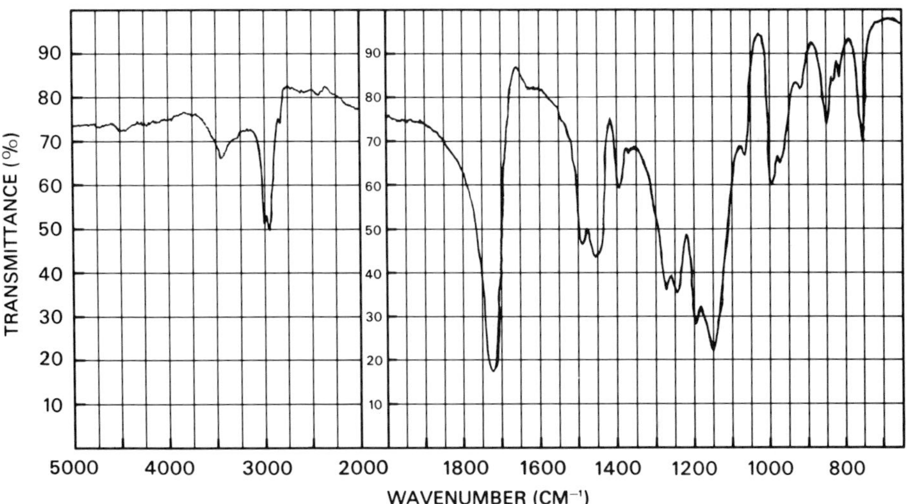

Fig. 5.12

(a) polymerised alkene

$$\ldots\underset{}{\overset{R}{\overset{|}{CH}}}-\overset{R}{\overset{|}{CH}}-\overset{R}{\overset{|}{CH}}-\ldots$$

(b) polyamide $\ldots NH-CO-(CH_2)_n-CO-NH-(CH_2)_n-NH\ldots$

(c) polyacrylate

$$\ldots CH_2-\overset{COOR}{\overset{|}{\underset{R'}{\underset{|}{C}}}}-CH_2-\overset{COOR}{\overset{|}{\underset{R'}{\underset{|}{C}}}}-\ldots$$

Continued overleaf

(d) phenolformaldehyde

OH OH OH

$-CH_2-$ CH_2OH

CH_2

(e) polyurethane $\ldots-R-NH-\underset{\underset{O}{\|}}{C}-O-R-O-\underset{\underset{O}{\|}}{C}-NH-R-\ldots$

(f) terylene (polyterephthalic ester)

$-\underset{\underset{O}{\|}}{C}-C_6H_4-\underset{\underset{O}{\|}}{C}-O-CH_2-CH_2-O\underset{\underset{O}{\|}}{C}-C_6H_4-\underset{\underset{O}{\|}}{C}-\ldots$

Problem 10
F336 Classify the plastic whose spectrum is shown in Fig. 5.13.

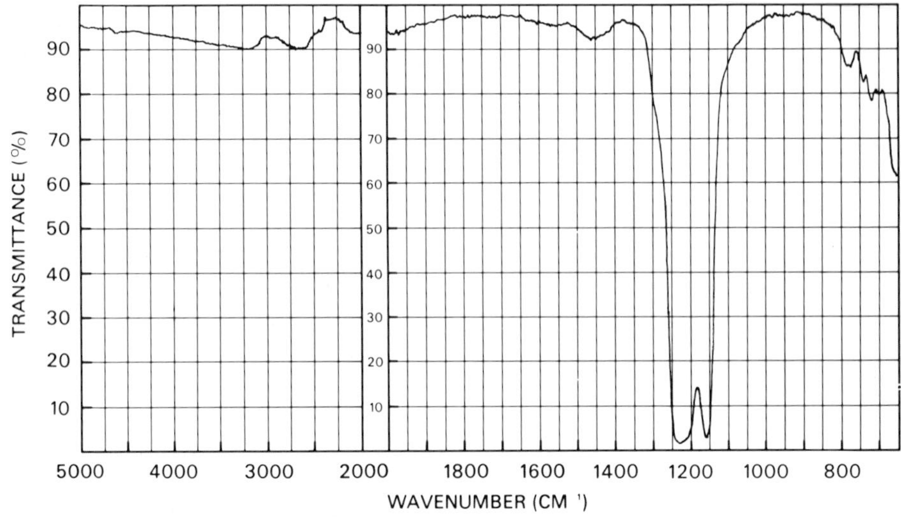

Fig. 5.13

Problem 11
F337 When acetone was passed through a length of PVC tubing and the effluent evaporated on a steam bath as far as possible a syrupy liquid was obtained. This yielded the spectrum shown in Fig. 5.14. What can you say about this material?

A335 (c). Note the C=O band at 1730 cm^{-1} in the normal ester position and the absence of NH_2, OH and aromatic bands.

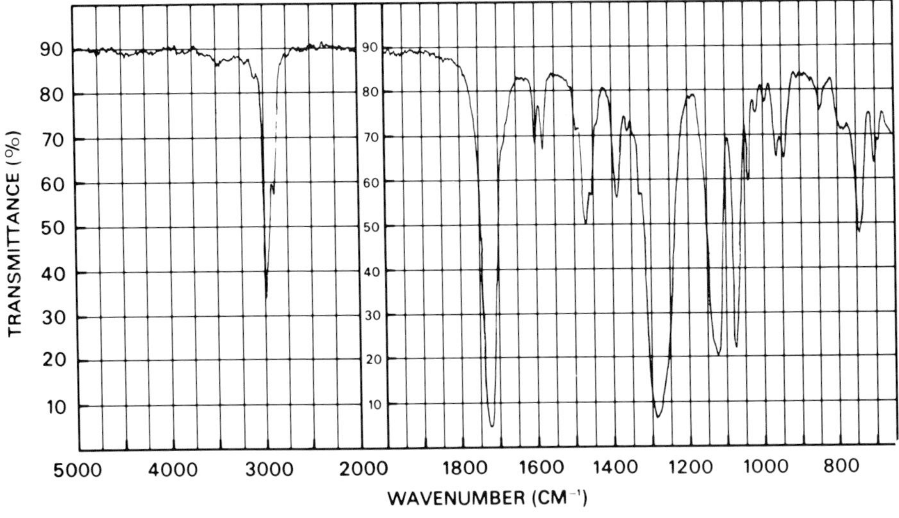

Fig. 5.14

A336 It contains no hydrogen and hence, provided it is not inorganic (which is unlikely from the sharpness of the bands and the absence of absorption near 1000 cm^{-1} due to Si—O etc.), it must be a perhalogenated material. The strong absorption near 1200 cm^{-1} indicates that fluorine is the halogen and the absence of strong absorption near 750 cm^{-1} suggests that little (if any) chlorine is present. The compound is in fact polytetrafluoroethylene (Teflon).

A337 The presence of the carbonyl band and the strong band at 1200 cm^{-1} suggest that the compound is an ester. The low frequency of the carbonyl band (for an ester) suggests that the ester C=O group is conjugated (to C=C or an aryl group). The bands near 1500 cm^{-1} and 1600 cm^{-1} also support the presence of a benzenoid ring which, from the appearance of the 700—900 cm^{-1} region, is probably 1,2-disubstituted. (Monosubstitution usually gives two bands of comparable intensity). Strong absorption just below 3000 cm^{-1} indicates that there is also a significant aliphatic residue in the molecule.

Although the spectrum does not distinguish between

COOR
R′

and

COOR
COOR

it is well-known that phthalic esters are commonly used as plasticizers which makes the latter far more probable. The compound is in fact dibutyl phthalate.

Conclusion

Most chemists will agree that competence in interpretive spectroscopy comes not only from a knowledge and understanding of the principles involved but also from experience in handling spectra. This leads to a systematic approach to a structural problem backed up by an awareness of sources of data already available. We have attempted in this programme to teach the principles of infrared spectroscopy and to indicate how they can be used to derive structural information from infrared spectra. From Part 3 onwards we have stressed the value of empirical information and this should have become even clearer to you as you tackled the problems. Although the use of empirical information in the form of tables of band frequencies will often lead to the solution of a problem, in many instances an impression gained from a complete spectrum when compared with reference spectra of known compounds will be more informative. Some useful compilations of infrared spectroscopic data are listed in the further reading for Part 5.

The most instructive exercises are undoubtedly those concerned with the work in your own laboratory. If you wish to supplement this, however, some stimulating problems in infrared spectroscopy are to be found in the books listed in the first part of the further reading section. In many of these, as with most of the problems you will meet in the laboratory, the infrared spectrum will be only a part of the information available. Once you have tackled a few problems of this sort you will realise that some ways of integrating the use of infrared data with that from other sources is more effective than others. For example, it is usually less rewarding to attempt to obtain all the information from the infrared spectrum in one operation than to keep returning to it as the structural possibilities are narrowed down. Finally, you should appreciate that the programme is an **introduction** to the use of infrared spectroscopy and as such it covers only the most commonly occurring correlations.

We hope that by using the format of a linear programme we have been able to give you some useful experience in interpretation of spectra. So often the subject matter of a theoretical treatise on the technique seems to fall short of the requirements for solving one's own particular problems. We hope that this has not been the case with this book.

Further Reading

(a) Problems

K. Nakanishi, *Infrared Absorption Spectroscopy—Practical*, Holden and Day, New York, 1962.

A. J. Baker, T. Cairns, G. Eglinton and F. J. Preston, *More Spectroscopic Problems in Organic Chemistry*, 2nd Edn., Heyden and Son, London, 1975.

R. M. Silverstein and G. C. Bassler, *Spectrometric Identification of Organic Compounds*, John Wiley and Sons, New York, 1967.

M. B. Winstead, *Organic Chemistry Structural Problems*, Heyden and Sons, London, 1969.

D. J. Pasto and C. R. Johnson, *Organic Structure Determinations*, Prentice-Hall, Englewood Cliffs, N.J., 1969.

J. D. Roberts and M. J. Caserio, *Organic Chemistry Problems*, Benjamin, New York, 1967.

B. Trost, *Problems in Spectroscopy*, Benjamin, New York, 1967.

D. H. Williams and I. Fleming, *Spectroscopic Problems in Organic Chemistry*, McGraw-Hill, New York, 1967.

A. Ault, *Problems in Organic Structure Determination*, McGraw-Hill, New York, 1967.
T. Goto, Y. Hirata and G. H. Stout, *Seminar Problems in Organic Chemistry*, Holden and Day, New York, 1967.
F. Scheinmann, *An Introduction to Spectroscopic Methods for the Identification of Organic Compounds*. Volume 1, Pergamon Press, Oxford, 1970.
J. R. Dyer, *Spectroscopic Problems in Organic Chemistry*, Prentice-Hall, Englewood Cliffs, N.J., 1972.
B. J. Brisden and D. W. Brown, *Spectroscopic Problems in Chemistry*, Van Nostrand, New York, 1973.

Most of the above books contain ultraviolet, n.m.r. and mass spectral data in addition to infrared spectra.

(b) Collections of spectra

Sadtler Standard Spectra, Sadtler Research Laboratories Inc., Philadelphia, (1) *Infrared Prism Standard Spectra*, 45 000 spectra in 45 volumes; (2) *Infrared Grating Spectra*, 30 000 spectra in 30 volumes.

Inc., Philadelphia, 9000 spectra in 9 volumes.
R. Mecke and F. Langenbucher, *Infrared Spectra of Selected Chemical Compounds*, Heyden and Son, London, over 1800 spectra in chemical classes order with indexes; available only on microfilm and microfiche.

Alphabetical List of Compound Names, Formulae and References to Published Infrared Spectra—AMD 34, A.S.T.M., Philadelphia; contains over 92 000 references of published literature of infrared spectra, 1969.

AMD 34–S14, A.S.T.M., Philadelphia; 14th supplement to AMD 34 containing references to 10 000 compounds, 1972.

Selected Values of Properties of Hydrocarbons and Related Compounds, The American Petroleum Institute Research Project 44 (API RP 44), 3242 spectra available, 1974.

Selected Values of Properties of Chemical Compounds, The Thermodynamics Research Center Data Project (TRC), 901 spectra available, 1974.

C. J. Pouchert, *The Aldrich Library of Infrared Spectra*, Aldrich Chemical Co. Inc., Milwaukee, Wisconsin, 8000 spectra selected from the Aldrich company library, 1970.

R. G. J. Miller and H. A. Willis, *IRSCOT—Infrared Structural Correlation Tables and Data Cards*, Heyden and Son, London.

Answers to Multiple Choice Tests

Part 1	Part 3	Part 4
1. (c)	1. (b)	1. (a)
2. (b)	2. (b)	2. (b)
3. (a)	3. (c)	3. (c)
4. (b)	4. (b)	4. (a)
5. (c)	5. (c)	5. (c)
6. (b)	6. (a)	6. (a)
7. (a)	7. (c)	7. (a)
8. (b)	8. (a)	8. (c)
9. (a)	9. (b)	9. (b)
10. (a)	10. (b)	10. (c)

List of Compounds

Compounds whose spectra are discussed in the programme are listed below. Where spectra are reproduced the page number is italicized.

N

P

S

T

X

Subject Index

E

F

G

H

I

K

L

M

N

O

P

R

S

T

V

W

X